Contents

A Lion in the Meadow

by Margaret Mahy

The little boy said, 'Mother, there is a lion in the meadow.'

The mother said, 'Nonsense, little boy.'

The little boy said, 'Mother, there is a big yellow lion in the meadow.'

The mother said, 'Nonsense, little boy.'

The little boy said, 'Mother, there is a big, roaring, yellow, whiskery lion in the meadow!'

The mother said, 'Little boy, you are making up stories again. There is nothing in the meadow but grass and trees. Go into the meadow and see for yourself.'

The little boy said, 'Mother, I'm scared to go into the meadow, because of the lion which is there.'

The mother said, 'Little boy, you are making up stories — so I will make up a story, too. . . . Do you see this match box? Take it out into the meadow and open it. In it will be a tiny dragon. The tiny dragon will grow into a big dragon. It will chase the lion away.'

The little boy took the match box and went away. The mother went on peeling the potatoes.

Suddenly the door opened.

In rushed a big, roaring, yellow, whiskery lion.

'Hide me!' it said. 'A dragon is after me!'

4

The lion hid in the broom cupboard.

Then the little boy came running in.

'Mother,' he said. 'That dragon grew too big. There is no lion in the meadow now. There is a DRAGON in the meadow.'

The little boy hid in the broom cupboard too.

'You should have left me alone,' said the lion. 'I eat only apples.'

'But there wasn't a real dragon,' said the mother. 'It was just a story I made up.'

'It turned out to be true after all,' said the little boy. 'You should have looked in the match box first.'

'That is how it is,' said the lion. 'Some stories are true, and some aren't. . . . But I have an idea. We will go and play in the meadow on the other side of the house. There is no dragon there.'

'I am glad we are friends now,' said the little boy.

The little boy and the big roaring yellow whiskery lion went to play in the other meadow. The dragon stayed where he was, and nobody minded.

The mother never ever made up a story again.

Maggy Scraggle Loves the Beautiful Ice-Cream Man

by Jill McDonald

Once there was a witch called Maggy Scraggle.

She was quite ugly but she never looked in a mirror so it didn't bother her much.

She could do a few simple magic spells — nothing fancy.

Even the simple spells didn't always turn out right.

She lived in a mucky hut among the bogs and boulders and in a gloomy sort of way she was happy enough.

At the top of the hill was a pretty palace.

That was where King Jinko and Queen Pozy lived, and also the Prince, the handsome Sintypuppa.

Though they were royal they were not rich — they were quite poor in fact. King Jinko worked hard to grow follyberries.

Queen Pozy worked hard making the follyberries into follyberry tarts to sell because they needed the money.

Sintypuppa was rather spoilt. He had a very posh motor scooter, and when he wanted to get tattooed all over, the King and Queen let him, even though it cost a lot.

But one evening the King said seriously, 'My son, I wish you would get married. The times are very boring and the people need a royal wedding to cheer them up. Also your Ma could do with a bit of help in the kitchen.'

'Oh, all right,' said Sintypuppa. 'I'll set off tomorrow at sun-up and find a pretty girl to be my wife.'

'But,' he thought to himself, 'she must not know I am the prince. She must marry me for love alone. I will disguise myself.'

He rummaged around in the attic for some fancy old finery and dressed himself up in it.

Queen Pozy gasped when she saw him.

'It's me, Sintypuppa,' said the Prince.

'Gracious,' she said, 'I never would have known you.'

'Now, dear Mother,' said the Prince, 'kindly make a hundred ice-creams for me to sell, and my disguise will be complete.'

'I never saw an ice-cream man looking like that,' murmured Queen Pozy.

The King and Queen worked all night making the follyberries into follyberry ice-creams and packing them into a little freezer cart which hooked on to Sintypuppa's motor scooter.

At sun-up Sintypuppa came down in all his finery, thanked the King and Queen for working so hard, started up his motor scooter, Vrooom, Vrooom, and set off to find himself a wife.

'All the girls will want to marry a gorgeous ice-cream man like me,' he thought as he zoomed down the hill when suddenly . . .

. . . Wham! Zock! The motor scooter crashed into a boulder; Sintypuppa flew through the air and . . .

. . . landed — Kersplog! — in the bog. All the ice-creams fell plop in the mud.

'Help, Help, Oh Help!' shouted Sintypuppa.

Maggy Scraggle rushed to her door.

'Oh, beautiful ice-cream man,' she cried, 'stay quite still and I'll get you out.'

'Aaaaaah! A rickety witch!' screamed Sintypuppa. He was so scared he scrabbled out all by himself and zoomed back home at top speed.

'Well, of all the cheek!' said Maggy Scraggle.

She scrubbed the dust off her mirror and had a look at herself.

'Good grief,' she gasped. 'I *am* a bit of a fright.'

She jumped on her broomstick and whisked away to town where she bought a few bits and pieces with which to prettify herself.

She sewed on buttons and beads and feather flowers. She sewed pretty patches over the plain patches. She put on a golden wig and long curly eyelashes.

She put a flashy ring on every finger. She painted her fingernails and even her toenails. She put on plenty of rouge and lipstick.

Then she went out to find the ice-cream man.

Down the hill he came again with his finery all washed and ironed and a batch of fresh-frozen ice-creams. At the sight of Maggy Scraggle he slammed on the brakes — Screeeech!

The motor scooter skidded to a stop.

'Oh, beautiful ice-cream man,' said the witch, 'don't be shy, don't make a fuss, just say you'll marry me.'

'What? A splendid fellow like me marry an ugly old witch?' shouted Sintypuppa.

Maggy Scraggle was hopping mad. 'After all the trouble I've taken to prettify myself!' she squawked. 'Why, you saucy young whippersnapper!'

She flickered her fingers, spun around widdershins, muttered some magic words and — Karamba! — Sintypuppa was turned into a . . .

. . . frog!

'Serves him right!' snorted Maggy Scraggle. And she stomped back home again.

'Oh!' pitifully croaked Sintypuppa as he hopped up the hill to the palace.

He couldn't talk to the King and Queen so he wrote them a letter to explain what had happened.

'Oh, my hapless little Sintypuppa,' sobbed Queen Pozy. King Jinko was wild.

'By golly, I'll give that witch a piece of my mind,' he shouted.

'Go and get her at once,' he ordered his two soldiers. (He couldn't afford more than two.)

Back they came in a trice with Maggy Scraggle all snappish and pettish.

'Now then, Madam,' thundered King Jinko. 'How dare you? Unmagic this prince *at once*, do you hear!'

Maggy Scraggle blenched. She was quite scared to learn it was a prince she had magicked into a frog.

She tried her hardest but it is not a simple spell that unmagics a frog. Even before it was finished she guessed it was going all wrong.

Suddenly, Maggy Scraggle disappeared and there were *two* frogs!

'Why, that stupid old faggot has turned *herself* into a frog!' yelled King Jinko.

Queen Pozy fainted.

10

Maggy Scraggle was surprised, not to say aghast, at what had happened. But Sintypuppa went down on his knees.

'Oh, dearest, delightful, Maggy Scraggle,' he croaked, 'only say you will wed me and my joy will know no bounds.'

'Well, I might do worse,' thought Maggy Scraggle.

'I accept,' she croaked.

Together they wrote a letter to the King.

'Oh heavens,' groaned King Jinko. 'Whatever will the people say?' He poured some ice-cream down Queen Pozy's neck, which woke her up quite quickly.

He showed her the letter.

'Oh, Oh,' sobbed the Queen. Then suddenly she stopped, wiped her eyes and became sensible.

'We should announce the royal wedding without further fuss or ado,' she said.

So Prince Sintypuppa and Maggy Scraggle were wed.

As they were so tiny it didn't cost nearly as much as most royal weddings, which pleased King Jinko.

Maggy Scraggle and Prince Sintypuppa were married on the magnificent wedding cake itself. A red velvet carpet was laid so there wouldn't be footprints on the icing.

It was so unusual to have a royal wedding between two frogs that all the television channels wanted to have it on their programmes, and it was broadcast 'live by satellite' all over the world. The television people paid King Jinko thousands of pounds, so the royal family became very rich.

The King and Queen gave Sintypuppa and Maggy Scraggle a tiny frog palace for their wedding present. It was built in a nice damp squishy place by the fountain and they lived there in great harmony and happiness for a very long time.

Pete and the Ladybird

by Leila Berg

One day, Pete found a matchbox. It was an empty matchbox, so Pete started to look for something to put in it.

He looked very hard all over the pavement. When he saw his shadow standing beside him, he said to his shadow, 'You look too.' So his shadow looked too.

A lady saw Pete bent over, looking. She stopped and asked him, 'Have you lost something?'

Pete straightened up. He shook his head. 'No, thank you,' said Pete. And he went on looking.

The lady was not in a hurry, so she stayed beside him. Pete looked all over the pavement. Then he started looking in the gutter. At last the lady said, 'Why are you looking, if you haven't lost anything?'

Pete turned his head. He said to her, 'I'm looking for something to put in my matchbox, because there's nothing in it.' And he went on looking, and took no notice of her at all, because he was very busy.

She was just going to go far away, because she could see how busy he was, when she saw a feather come floating down from the sky. She reached out her hand, and caught it.

'Look,' she said. And she gave it to Pete.

'It's part of an eagle,' he said. 'I mean a seagle.' And he nodded his head, and put the feather in his matchbox.

But the feather wouldn't fit. It was much too big. 'Bother,' said Pete. 'We'd better give it back,' he said.

He stared up into the sky with its white racing clouds. There were three or four birds flying round and round. One of them was crying. 'I think it's that one,' said Pete. 'He's looking for it.' And he put the feather on his hand, and blew it. It sailed up, up into the air, and up into the sky. It was a very windy day. 'I hope the seagle caught it,' said Pete. 'It was his tail, I expect.' And he went on looking.

Now he found a holly leaf. It was a baby holly leaf, with soft baby prickles that didn't hurt at all. Pete liked it because it was so little, and green and shiny, and because its prickles were soft and tickly, and didn't hurt. He tried it in the matchbox, and it was just right.

'Good,' said Pete. 'It's the same long.'

'I think you mean the same length,' said the lady.

'I know,' said Pete. 'That's what I said.' And he shut the lid, and put the box in his pocket.

Then the lady went away, because there was nothing else to help with. 'Goodbye,' she said, and Pete said goodbye, too.

The wind was blowing the clouds over the sun; one minute it was bright, one minute it was dull.

Pete began to play a game called Shadow-BANG. The way he played it was like this.

He stood on the pavement, with his legs wide apart. Then very suddenly, keeping his legs stiff, he bent down and looked between them and shouted 'Shadow-BANG'. And sometimes his shadow was there, and then Pete won. And sometimes his shadow wasn't, and then Pete didn't win *absolutely*, only almost.

Now, just as Pete had his head between his legs, and was opening his mouth to shout 'Shadow-BANG' for the nineteenth time, a ladybird came along the pavement. She was behind Pete, and

Pete saw her through his legs. She came nearer and nearer. And as she got nearer, Pete moved his head back through his legs again, till he was simply standing on the pavement with his hands on his knees, staring at a ladybird.

She was an orange-coloured ladybird with three big spots. 'Hello, ladybird,' Pete said, as she went past his shoe. But she didn't answer.

Pete very carefully kneeled down on the pavement, and put his finger in front of the ladybird. She thought about his finger. Pete could see her thinking. Then just as he thought she didn't like his finger, she decided that she did. She climbed on to it, and Pete slowly stood up.

Then he walked along carefully. He was taking the ladybird home.

He was looking so hard at the ladybird as he walked along carefully, that he bumped right into a window-cleaner. The window-cleaner was propping up a ladder against the wall.

'Look out!' he shouted. 'Where d'you think you're going?'

But Pete was staring at his finger. The ladybird had gone. She had opened her wings and flown straight into the wind. And now he couldn't see her anywhere.

He turned to the window-cleaner. He was very upset. 'You've lost my ladybird,' he said tearfully. 'I was taking her home and she was going to live in my matchbox, and now

14

she's gone and she'll get lost, and it's all your fault. . . .'

'Hush,' said the man, flapping his duster in front of Pete's face.

Pete was rather surprised at that, and he hushed.

'Were you telling me off about something?' said the window-cleaner.

'Yes, I am telling you off,' said Pete. 'I'm telling you off because you lost my ladybird and she'll get lost and I was taking her home—'

'Wait a minute,' said the window-cleaner. 'It's a funny thing you should be telling *me* off, because I was telling *you* off. What's the idea of bumping into me, and not even saying you were sorry? You might have sent this ladder right through the window, you know, and smashed all the glass and everything. If only you'll look where you're *going*,' he said. 'I'll look where you're *coming*.'

Pete laughed. Then the man said, 'Shall I tell you something?'

'What?' said Pete.

'I think your ladybird was glad to fly away. I don't mean she didn't like you,' he said quickly, because he could see Pete was going to get cross again. 'I expect she liked you very much, but really she wanted to fly about in the sun. She was just walking along with you because she liked you. And then it was time for her to go.'

'I know,' said Pete. 'It isn't time for me to go. Can I help you clean windows?'

'Well, just for a little bit,' said the window-cleaner.

So Pete stood on the second rung of the ladder, and cleaned a little piece of window, just as big as his head. And the window-cleaner did the rest because he was used to it. And when they had finished, all the blue sky shone in the window and the sun came and sat in it.

'Didn't we make it shiny?' said Pete.

That was a good day, too.

A List
by Arnold Lobel

One morning Toad sat in bed.

'I have many things to do,' he said.

'I will write them all down on a list so that I can remember them.'

Toad wrote on a piece of paper: A list of things to do today

Then he wrote: Wake up

'I have done that,' said Toad, and he crossed out: ~~Wake up~~

Then Toad wrote other things on the paper.

'There,' said Toad. 'Now my day is all written down.'

He got out of bed and had something to eat.

Then Toad crossed out: ~~Eat breakfast~~

Toad took his clothes out of the cupboard and put them on.

Then he crossed out: ~~Get dressed~~

Toad put the list in his pocket.

He opened the door and walked out into the morning.

Soon Toad was at Frog's front door. He took the list from his pocket and crossed out: ~~Go to Frog's house~~

Toad knocked at the door.

'Hello,' said Frog.

① A list of things to do today

② Wake up

③ Eat breakfast

④ Get dressed

⑤ Go to frog's house

⑥ Take walk with frog

⑦ Go to sleep

'Look at my list of things to do,' said Toad.

'Oh,' said Frog, 'that is very nice.'

Toad said, 'My list tells me that we will go for a walk.'

'All right,' said Frog. 'I am ready.'

Frog and Toad went on a long walk. Then Toad took the list from his pocket again.

He crossed out: ~~Take walk with frog~~

Just then there was a strong wind. It blew the list out of Toad's hand! The list blew high up into the air.

'Help!' cried Toad. 'My list is blowing away. What will I do without my list?'

'Hurry!' said Frog. 'We will run and catch it.'

'No!' shouted Toad. 'I cannot do that.'

'Why not?' asked Frog.

'Because,' wailed Toad, 'running after my list is not one of the things that I wrote on my list of things to do!'

Frog ran after the list. He ran over hills and swamps, but the list blew on and on.

At last Frog came back to Toad.

'I am sorry,' gasped Frog, 'but I could not catch your list.'

'Blah,' said Toad. 'I cannot remember any of the things that were on my list of things to do. I will just have to sit here and do nothing,' said Toad.

Toad sat and did nothing.

Frog sat with him.

After a long time Frog said, 'Toad, it is getting dark. We should be going to sleep now.'

'Go to sleep!' shouted Toad. 'That was the last thing on my list!'

Toad wrote on the ground with a stick: go to sleep

Then he crossed out: ~~go to sleep~~

'There,' said Toad. 'Now my day is all crossed out!'

'I am glad,' said Frog.

Then Frog and Toad went to sleep.

The Fish Cart
by Ruth Manning-Sanders

Fox crept under some bushes by the roadside. He was *hungry*. He hadn't eaten anything since yesterday at dinner-time, and now it was dinner-time again. *Sniff, sniff, sniff* went Fox's nose, poking among the bushes.

'We're wasting our time here,' said Fox's nose. 'There's been a rabbit, but he's gone.'

So Fox crept out from under the bushes on to the edge of the road. '*Sniff, sniff, sniff!* Oh, oh, now I *do* smell something!' whispered Fox's nose. 'I smell something delicious! I smell *fish*!'

And see, coming along the highroad towards Fox, a man driving a cart piled high with fish!

'My dinner, my dinner!' thought Fox, backing under the bushes again. 'But how to get it?' Yes, indeed, how to get it?

The cart trundled slowly past. It was drawn by an old grey horse and driven by a man with a red face, who sat on one of the shafts roaring out a song.

'It's a low cart, we could easily spring up into it,' Fox's legs told him.

'And eat and eat and *eat*,' said Fox. . . . 'But no, the man may have a gun, and then I should be dead before I'd swallowed a mouthful! . . . Dead, eh? Well now, that *is* an idea!'

19

So, as Fox sat thinking, who should come bounding over the hedge but Wolf. Wolf was hungry too; he was making an angry face and gnashing his teeth. 'Cousin Fox, Cousin Fox,' he snarled, 'find me something to eat this *minute* — or I shall eat *you*!'

Fox knew he was much cleverer than Wolf, so he wasn't frightened. 'Just keep calm, Cousin Wolf,' said he. 'Wait here, and I'll get you food — plenty of it.'

The fish cart had disappeared round a bend of the road, but it was going slowly. It wasn't very far ahead, so Fox's ears told him. They could still hear the rumble of the cartwheels, and the driver's loud song.

'Off with us then!' said Fox's legs, and they pushed through the hedge into a field, galloped faster than fast across the field, and came out again on to the highroad ahead of the fish cart. Then Fox lay down with his eyes shut, and his body stiff, and his legs sprawled out. He was pretending to be dead.

Along came the cart. The big man with the red face pulled up the old grey horse. What was that lying there? A dead fox, eh! And he jumped down from the shaft he was sitting on, picked up Fox, and tossed him into the cart on top of the fish. Then he scrambled on to the shaft again, and drove on.

'*Troll-a-loll, troll-a-loll!*' The man was roaring out a song about how he would skin Fox, and how his wife would make him a warm winter waistcoat of fox fur. His roaring voice, the rumble of the wheels, and the *clop, clop* of the horse's hoofs clattered about Fox's ears as he lay stretched out stiffly on top of the pile of fish pretending to be dead. . . .

But what was Fox doing now? He was quickly, quickly moving his feet and kicking one fish after another out of the cart and down on to the road.

But what was Wolf doing? Well, Wolf had got tired of waiting. He had a smell of fish in his nostrils, and the smell was coming from somewhere along the highroad in front of him. So away trotted Wolf, and he had just got round the bend of the road when — did you ever? — fish, *fish*, FISH, a strew of fish scattered ahead of him!

'Oh, my dinner, my dinner, my good dinner!' Wolf opened his mouth as wide as it would go, and began gulping down one fish after another. And on he went, and on he went gulping and gobbling. . . .

The man sitting on the shaft of the cart was still singing about his Sunday waistcoat of fox fur. He didn't look behind him, no, not once. Fox pushed the last fish out of the cart, opened his eyes, gave a leap down on to the road and saw Wolf looking fat and pleased with himself, swallowing down the very last fish.

'Oh, Cousin Wolf, Cousin Wolf, you big greedy thief! There was enough fish for us both, more than enough for us both, and you've eaten them all!'

'Of course I've eaten them all,' said Wolf. 'You promised to get me plenty of food. You didn't say you wanted any yourself. And I'm really grateful to you, Cousin Fox,' said he, licking his lips and smiling fatly.

Fox was so angry that he flew at Wolf, thinking to bite him. But Wolf lifted his lip and snarled, so Fox thought better of it. Wolf was three times as big as he was.

'There are other ways!' he muttered.

Then he turned and trotted off home, with his poor empty stomach making grumbling noises, and his busy brain making plans to get his own back on greedy Wolf.

The Thrush Girl

by Godfried Bomans

Once upon a time there was a little girl who longed to be able to understand animals. She went to her old grandmother and said, 'Oh, grandmother, I would so love to understand the animals. Can you teach me how to do it?'

The grandmother could do a little magic, but not much.

'Oh, my child,' she said, 'I can only understand the thrushes, and that is not worth the trouble.'

'It is enough for me,' replied the little girl. 'Please teach me.'

So the grandmother taught the little girl to understand the thrushes. It was much easier than she had thought. She had only to be kind to the birds and throw them a few crumbs from time to time.

And when she had learned, she walked through the woods listening to the thrushes. Her grandmother was right. They had not much to say. But the little girl was right, too, because it was enough for her. And she went to her father and said, 'Father, bring in the hay, for it will rain tomorrow.'

The father believed the child. He brought in the hay and by evening it was stacked in the barn. The next morning it began to rain. All the hay in the district was soaked and only his was dry. The father was glad he had listened to his child, but all the other farmers were angry. They were not pleased about the hay that had been

23

saved; they thought only of their own loss. 'That girl is no good,' they said, 'she will go to the bad.'

But the little girl said nothing. She went back to her grandmother and said, 'Now I would like to understand the moles as well.'

'Oh, my child,' replied the grandmother, 'what a lot you want. Just be good to the animals, then you will understand them after a time.'

It was three months before the little girl could understand the moles, too. And one day she said to her father, 'Bury the potatoes deep in the ground, for tomorrow it will freeze.'

And sure enough, next day it froze. The father had buried his potatoes deep in the ground and they were undamaged. All the other farmers moaned and groaned, for they had not a single good potato left. He had warned them all, but they had not listened to him. They thought the little girl was bewitched. The little girl said nothing. Now she wanted to understand the language of the bees, too. Her grandmother could not help her, for she herself could not understand the bees.

'You know more than I now,' she said. 'You must learn to talk with bees in your own way.'

So the little girl was very kind to the bees. She no longer ate honey, but left what there was in the hives. And after a little while she could understand exactly what the bees were saying to each other. And one day she went to her mother and said, 'Prop up the fruit trees in the orchard and lock the windows tight, for there will be a great storm tonight.'

And that very evening a mighty wind rose up and devastated all the houses. The trees bowed and broke and there was great distress in the land. Only the orchard in which the little girl lived stood upright and not a tree was harmed. Other people grew so angry that they said, 'The child is bewitched! We will burn her!'

The people came running from every side with dry branches to build a great fire. The little girl stood on top of the pile of branches

24

and called in a loud voice, 'Thrushes, thrushes, help me now!'

The people could not understand what she was saying, for it was in bird language, but the thrushes understood her perfectly. And in their thousands they flew down and each plucked a twig from the fire. The little girl was soon standing on the grass and there was no more firewood to be seen. She ran happily home and cried, 'Father, mother, the birds have set me free!'

Oh, how happy they were! But their joy did not last long, for the king's soldiers knocked at the door.

'Open!' they cried. 'We have come to fetch the girl!'

She was taken to the market-place and there stood a man with a great gleaming axe.

'Your head must come off,' he said. 'Kneel down and stretch out your neck.'

The little girl did as she was told, but as she laid her head on the block she cried in a loud voice, 'Bees, bees, help me now!'

The executioner did not understand her, for she spoke in bee language. But the bees understood her perfectly. And just as the executioner raised his axe, there came a loud humming and thousands of bees dived at him with their stings ready to strike. The executioner fell to the ground, dead, and the king's soldiers scattered in terror. The little girl ran home at top speed and cried, 'Father, mother, here I am! The bees have saved me!'

But now the king himself took a hand. He rang the bell and said, 'I have come to fetch your daughter. I cannot kill her. But she shall be shut up in a stone tower.'

So the girl was taken to a tower with walls fifteen feet thick. The windows had iron bars and the door was shut with three locks. It was so dark that she could see nothing, but she heard the rats and mice scuttling across the stone floor. The little girl sat on the ground and began to weep bitterly.

'I shall never get out,' she said. 'Oh, moles, moles, help me now!'

No sooner had she said this than
thousands of moles began to tunnel
under the tower. The walls began to tilt
and topple. The floor burst open and the
bars sprang out of the window-frames.
And suddenly, crash! The tower fell with
a roar that reverberated throughout the
country. The king was just having
breakfast when he heard it. He put down
his knife and fork and said, 'The tower
has fallen. Did you hear it?'

'Yes, father,' said the prince. 'Shall I
marry her now?'

'I think you should,' said the king,
'although she is rather small.'

And together they stepped into their
coach with the four horses and rode at
full speed to the house where the little
girl lived. She was standing in the
garden, scattering crumbs for the birds.

'Will you marry me, later on?' asked
the prince.

'No,' said the little girl, 'I will not. I do
not like the people here. I shall go away.'

She packed three jam sandwiches in
her basket and added a few blackcurrants.
Then she put on her fur-trimmed cloak
and went away. And no one ever saw
her again.

The Giant who Threw Tantrums

by David Harrison

At the foot of Thistle Mountain lay a village.

In the village lived a little boy who liked to go walking. One Saturday afternoon he was walking in the woods when he was startled by a terrible noise.

He scrambled quickly behind a bush.

Before long a huge giant came stamping down the path.

He looked upset.

'Tanglebangled ringlepox!' the giant bellowed. He banged his head against a tree until the leaves shook off like snowflakes.

'Franglewhangled whippersnack!' the giant roared. Yanking up the tree, he whirled it around his head and knocked down twenty-seven other trees.

Muttering to himself, he stalked up the path towards the top of Thistle Mountain.

The little boy hurried home.

'I just saw a giant throwing a tantrum!' he told everyone in the village. They only smiled.

'There's no such thing as a giant,' the mayor assured him.

'He knocked down twenty-seven trees,' said the little boy.

'Must have been a tornado,' the weatherman said with a nod. 'Happens around here all the time.'

The next Saturday afternoon the little boy again went walking. Before long he heard a horrible noise. Quick as lightning, he slipped behind a tree.

Soon the same giant came storming down the path. He still looked upset.

'Pollywogging frizzelsnatch!' he yelled. Throwing himself down, he pounded the ground with both fists.

Boulders bounced like hailstones.

Scowling, the giant puckered his lips into an 'O'.

He drew in his breath sharply. It sounded like somebody slurping soup.

'Pooh!' he cried.

Grabbing his left foot with both hands, the giant hopped on his right foot up the path towards the top of Thistle Mountain.

The little boy hurried home.

'That giant's at it again,' he told everyone. 'He threw such a tantrum that the ground trembled!'

'Must have been an earthquake,' the police chief said. 'Happens around here sometimes.'

The next Saturday afternoon the little boy again went walking. Before long he heard a frightening noise.

He dropped down behind a rock.

Soon the giant came fuming down the path. When he reached the little boy's rock, he puckered his lips into an 'O'. He drew in his breath sharply with a loud, rushing-wind sound. 'Phooey!' he cried. 'I *never* get it right!'

The giant held his breath until his face turned blue and his eyes rolled up. 'Fozzlehumper backawacket!' he panted.

Then he lumbered up the path towards the top of Thistle Mountain.

The little boy followed him. Up and up and up he climbed to the very top of Thistle Mountain.

There he discovered a huge cave. A surprising sound was coming from it. The giant was crying!

'All I want is to whistle,' he sighed through his tears. 'But every time I try, it comes out wrong!'

The little boy had just learned to whistle. He knew how hard it could be. He stepped inside the cave.

The giant looked surprised. 'How did *you* get here?'

'I know what you're doing wrong,' the little boy said.

When the giant heard that, he leaned down and put his hands on his knees.

'Tell me at once!' he begged.

'You have to stop throwing tantrums,' the little boy told him.

'I promise!' said the giant, who didn't want anyone to think he had poor manners.

'Pucker your lips . . .' the little boy said.

'I always do!' the giant assured him.

'Then blow,' the little boy added.

'Blow?'

'Blow.'

The giant looked as if he didn't believe it. He puckered his lips into an 'O'. He blew. Out came a long, low whistle. It sounded like a railway engine. The giant smiled.

He shouted, 'I whistled! Did you hear that? I whistled!'

Taking the little boy's hand, he danced in a circle.

'You're a good friend,' the giant said.

'Thank you,' said the little boy. 'Perhaps some time we can whistle together. But just now I have to go. It's my suppertime.'

The giant stood before his cave and waved goodbye.

The little boy seldom saw the giant after that. But the giant kept his promise about not throwing tantrums.

'We never have earthquakes,' the mayor liked to say.

'Haven't had a tornado in ages,' the weatherman would add.

Now and then they heard a long, low whistle somewhere in the distance.

'Must be a train,' the police chief would say.

But the little boy knew his friend the giant was walking up the path towards the top of Thistle Mountain — whistling.

The Knee-High Man

by Julius Lester

Once upon a time there was a knee-high man. He was no taller than a person's knees. Because he was so short, he was very unhappy. He wanted to be big like everybody else.

One day he decided to ask the biggest animal he could find how he could get big. So he went to see Mr Horse. 'Mr Horse, how can I get big like you?'

Mr Horse said, 'Well, eat a whole lot of corn. Then run around a lot. After a while you'll be as big as me.'

The knee-high man did just that. He ate so much corn that his stomach hurt. Then he ran and ran and ran until his legs hurt. But he didn't get any bigger. So he decided that Mr Horse had told him something wrong. He decided to go ask Mr Bull.

'Mr Bull? How can I get big like you?'

Mr Bull said, 'Eat a whole lot of grass. Then bellow and bellow as loud as you can. The first thing you know, you'll be as big as me.'

So the knee-high man ate a whole field of grass. That made his stomach hurt. He bellowed and bellowed and bellowed all day and all night. That made his throat hurt. But he didn't get any bigger. So he decided that Mr Bull was all wrong too.

Now he didn't know anyone else to ask. One night he heard Mr Hoot Owl hooting, and he remembered that Mr Owl knew everything.

'Mr Owl? How can I get big like Mr Horse and Mr Bull?'

'What do you want to be big for?' Mr Hoot Owl asked.

'I want to be big so that when I get into a fight, I can whip everybody,' the knee-high man said.

Mr Hoot Owl hooted. 'Anybody ever try to pick a fight with you?'

The knee-high man thought a minute. 'Well, now that you mention it, nobody ever did try to start a fight with me.'

Mr Owl said, 'Well, you don't have any reason to fight. Therefore, you don't have any reason to be bigger than you are.'

'But, Mr Owl,' the knee-high man said, 'I want to be big so I can see far into the distance.'

Mr Hoot Owl hooted. 'If you climb a tall tree, you can see into the distance from the top.'

The knee-high man was quiet for a minute. 'Well, I hadn't thought of that.'

Mr Hoot Owl hooted again. 'And that's what's wrong, Mr Knee-High Man. You hadn't done any thinking at all. I'm smaller than you, and you don't see me worrying about being big. Mr Knee-High Man, you wanted something that you didn't need.'

My Naughty Little Sister Makes a Bottle-Tree

by Dorothy Edwards

One day, when I was a little girl, and my naughty little sister was a littler girl, my naughty little sister got up very early one morning, and while my mother was cooking the breakfast, my naughty little sister went quietly, quietly out of the kitchen door, and quietly, quietly up the garden-path. Do you know why she went *quietly* like that? It was because she was *up to mischief*.

She didn't stop to look at the flowers, or the marrows or the runner-beans, and she didn't put her fingers in the water-tub. No! She went right along to the tool-shed to find a trowel. You know what trowels are, of course, but my naughty little sister didn't. She called the trowel 'a digger'.

'Where is the digger?' said my naughty little sister to herself.

Well, she found the trowel, and she took it down the garden until she came to a very nice place in the big flower-bed. Then she stopped and began to dig and dig with the trowel, which you know was a most naughty thing to do, because of all the little baby seeds that are waiting to come up in flower-beds sometimes.

Shall I tell you why my naughty little sister dug that hole? All right, I will. It was because she wanted to plant a brown shiny acorn. So, when she had made a really nice deep hole, she put the acorn in it, and covered it all up again with earth, until the brown shiny acorn was all gone.

Then my naughty little sister got a stone, and a leaf, and a stick, and she put them on top of the hole, so that she could remember where the acorn was, and then she went indoors to have her hands washed for breakfast. She didn't tell me, or my mother or anyone about the acorn. She kept it for her secret.

Well now, my naughty little sister kept going down the garden all that day, to look at the stone, the leaf and the stick, on top of her acorn-hole, and my naughty little sister smiled and smiled to herself because she knew that there was a brown shiny acorn under the earth.

But when my father came home, he was very cross. He said, 'Who's been digging in my flowerbed?'

And my little sister said, 'I have.'

Then my father said, 'You are a bad child. You've disturbed all the little baby seeds!'

And my naughty little sister said, 'I don't care about the little baby seeds, I want a home for my brown shiny acorn.'

So my father said, 'Well, *I* care about the little baby seeds myself, so I shall dig your acorn up for you, and you must find another home for it,' and he dug it up for her at once, and my naughty little sister tried all over the garden to find a new place for her acorn.

But there were beans and marrows and potatoes and lettuce and tomatoes and roses and spinach and radishes, and no room at all for the acorn, so my naughty little sister grew crosser and crosser and when tea-time came she wouldn't eat her tea. Aren't you glad you don't show off like that?

Then my mother said, 'Now don't be miserable. Eat up your tea and you shall help me to plant your acorn in a bottleful of water.'

So my naughty little sister ate her tea after all, and then my mother, who was a clever woman, filled a bottle with water, and showed my naughty little sister how to put the acorn in the top of the bottle.

Shall I tell you how she did it, in case you want to try?

Well now, my naughty little sister put the pointy end of the acorn into the water, and left the bottom of the acorn sticking out of the top — (the bottom end, you know, is the end that sits in the little cup when it's on the tree).

'Now,' said my mother, 'you can watch its little root grow in the water.'

My naughty little sister had to put her acorn in lots of bottles of water, because the bottles were always getting broken, as she put them in such funny places. She put them on the kitchen window-sill where the cat walked, and on the side of the bath, and inside the bookcase, until my mother said, 'We'll put it on top of the cupboard, and I will get it down for you to see every morning after breakfast.'

Then at last, the little root began to grow. It pushed down, down into the bottle of water and it made lots of other little roots that looked just like whitey fingers, and my little sister was pleased as pleased. Then, one day, a little shoot came out of the top of the acorn, and broke all the browny outside off, and on this little shoot were

two little baby leaves, and the baby leaves grew and grew, and my mother said, 'That little shoot will be a big tree one day.'

My naughty little sister was very pleased. When she was pleased she danced and danced, so you can just guess how she danced to think of her acorn growing into a tree.

'Oh,' she said, 'when it's a tree we can put a swing on it, and I can swing indoors on my very own tree.'

But my mother said, 'Oh, no. I'm afraid it won't like being indoors very much now, it will want to grow out under the sky.'

Then my little sister had a good idea. And now, this is a *good thing* about my little sister — she had a *very kind thought* about her little tree. She said, 'I know! When we go for a walk we'll take my bottle-tree and the digger' (which, of course, you call a trowel) 'and we will plant it in the park, just where there are no trees, so it can grow and grow and spread and spread into a big tree.'

And that is just what she did do. Carefully, carefully, she took her bottle-tree out of the bottle, and put it in her little basket, and then we all went out to the park. And when my little sister had found a good place for her little bottle-tree, she dug a nice deep hole for it, and then she put her tree into the hole, and gently, gently put the earth all round its roots, until only the leaves and stem were showing, and when she'd planted it in, my mother showed her how to pat the earth with the trowel.

Then at last the little tree was in the kind of place it really liked, and my little sister had planted it all by herself.

Now you will be pleased to hear that the little bottle-tree grew and grew and now it's quite a big tree. Taller than my naughty little sister, and she's quite a big lady nowadays.

The Laughing Dragon

by Richard Wilson

There was once a King who had a very loud voice, and three sons.

His voice was *very* loud. It was so loud that when he spoke every one jumped. So they called the country he ruled over by the name of Jumpy.

But one day the King spoke in a very low voice indeed. And all the people ran about and said, 'The King is going to die.'

He *was* going to die, and he *did* die. But before he died he called his three sons to his bedside. He gave one half of Jumpy to the eldest son; and he gave the other half to the second son. Then he said to the third, 'You shall have six shillings and eightpence farthing and the small bag in my private box.'

In due time the third son got his six shillings and eightpence farthing, and put it safely away into his purse.

Then he got the bag from the King's private box. It was a small bag made of kid, and was tied with a string.

The third son, whose name, by the way, was Tumpy, untied the string and looked into the bag. It had nothing in it but a very queer smell. Tumpy sniffed and then he sneezed. Then he laughed, and laughed, and laughed again without in the least knowing what he was laughing at.

'I shall never stop laughing,' he said to himself. But he did, after

half an hour and two minutes exactly. Then he smiled for three minutes and a half exactly again.

After that he looked very happy; and he kept on looking so happy that people called him Happy Tumpy, or H.T. for short.

Next day H.T. set out to seek his fortune. He had tied up the bag again and put it into the very middle of his bundle.

His mother gave him some bread and a piece of cheese, two apples and a banana. Then he set out with a happy face. He whistled as he went along with his bundle on a stick over his shoulder.

After a time he was tired, and sat down on a large milestone. As he was eating an apple, a grey cat came along. It rubbed its side against the large stone, and H.T. stroked its head.

Then it sniffed at the bundle that lay on the grass. Next it sneezed, and then it began to laugh. It looked so funny that H.T. began to laugh too.

'You must come with me, puss,' said H.T. The cat was now smiling broadly. It looked up at H.T. and he fed it. Then they went on side by side.

By-and-by H.T. and the cat came to a town, and met a tall, thin man. 'Hallo,' he said, and H.T. said the same.

'Where are you going?' asked the man.

'To seek my fortune,' said H.T.

'I would give a small fortune to the man who could make me laugh.'

'Why?' said H.T.

'Because I want to be fat,' said the man, 'and people always say "laugh and grow fat." '

'How much will you give?' said H.T.

'Oh, five shillings and twopence halfpenny anyhow,' said the man.

H.T. put down his bundle and took out his bag. He held it up near the man's face and untied the string. The man sniffed and then he sneezed. Then he laughed for half an hour and two minutes. Next he smiled for three minutes and a half.

By that time he was quite fat. So he paid H.T. five shillings and twopence halfpenny. Then he went on his way with a smile and a wave of the hand.

'That is good,' said H.T. 'If I go on like this I shall soon make my fortune.' He tied up his bag and went on again. The grey cat walked after him with a smile on its face that never came off.

After an hour the two companions came to another town. There were a lot of men in the street, but no women, or boys, or girls. The men looked much afraid. H.T. went up to one of them, 'Why do you look so much afraid?' he asked politely.

'You will look afraid too, very soon,' said the man. 'The great dragon is coming again. It comes to the town each day, and takes a man and a cheese. In ten minutes it will be here.'

'Why don't you fight it?' asked H.T.

'It is too big and fierce,' said the man. 'If any man could kill it he would make his fortune.'

'How is that?' said H.T.

'Well,' said the man, 'the King would give him a bag of gold, and make the princess marry him.'

All at once H.T. heard a loud shout.

'The dragon is coming!' called a man who wore a butcher's apron. Then he ran into his shop, banged the door, and threw a large piece of meat out of the window. There was now nothing in the street but H.T., the cat, and the piece of meat.

H.T. did not run away, not even when he saw the huge dragon come lumbering up the street on all fours. It crept along, and turned its head this way and that. Its face had a terrible look.

Fire came out of its nose when it blew out. And three of the houses began to burn. Then it came to the meat. It sniffed it and stopped to

eat it. That gave H.T. time for carrying out his plan.

He took out his bag and untied the string. Then he threw it down before the dragon. On it came, blowing more fire from its nostrils. Soon the butcher's shop was burning. There was a noise like the noise from an oven when the meat is roasting.

The dragon still came on. When it got up to the bag it stopped. It sniffed. Then it sneezed so hard that two houses fell down flat. Next it began to laugh, and the noise was so loud that the church steeple fell into the street.

Of course it had stopped to laugh. It sat up on its hind legs and held its sides with its forepaws. Then it began to smile. And a dragon's smile, you must understand, is about six feet wide!

The dragon looked so jolly that H.T. did not feel afraid of it any more; not in the least. He went up to it and took one of its forepaws into his arm. The cat jumped on the dragon's head. And they all went along the street as jolly as sandboys.

A woman popped her head out of a high window. 'Take the first to the right,' she said, 'and the second to the left. Then you will come to the King's royal palace. You cannot miss it.'

'Thank you very much,' said H.T., and he and the dragon and the cat smiled up at her. H.T. waved his hand. The dragon waved its other forepaw. And the cat waved its tail.

So they went on — down one street and then another. At last they came to a big, open, green space in which stood a big palace. It had a wall round it with four large gates in it. At each gate there was a sentry box. But not one sentry could be seen.

H.T., with his friend the dragon, came smiling up to one of the gates. Above the gate H.T. saw someone peeping over the wall. 'He wears a crown,' he said to the dragon, 'so it must be the King.' The dragon kept on smiling.

'Hallo!' cried the King. 'What do *you* want?'

'Hallo!' cried H.T. 'I want the bag of gold and the princess.'

'But you have not killed the dragon,' said the King.

'I should think not,' said H.T. 'Why, he is my friend. He is my very dear friend. He will not do any harm now. Look at him.'

The King stood up and put his crown straight. It had fallen over one eye in his fright. The dragon went on smiling in a sleepy way. There was no fire in his nose now.

'But,' said the King, 'how do I know he will not begin to kill people again?'

'Well,' said H.T., 'we will make a big kennel for him and give him a silver chain. Each day I will give him a sniff from my empty bag. Then he will be happy all day and go to sleep every night.'

'Very well,' said the King. 'Here is the bag of gold. You will find the princess in the laundry. She always irons my collars. And you can have my crown as well. It is very hard and heavy. I do not want to be King any more. I only want to sit by the fire and have a pipe and play the gramophone.'

So he threw his crown down from the wall. The dragon caught it on his tail and put it on H.T.'s head. Then H.T. went to the laundry and married the princess right away.

And the dragon lived happy ever after, and so did the cat, and so did everybody else, at least until they died.

I ought to tell you that King H.T. used the bag all his life to keep the dragon laughing. He died at the age of 301 years, one month, a week, and two days.

The next day the dragon took a very hard sniff at the bag. And he laughed so much that he *died* of laughing.

So they gave the bag to the dentist. And when anyone had to have a tooth out he took a sniff. Then he laughed so much that he did not feel any pain. And when the tooth was out he was happy ever after, or at least until the next time he ate too many sweets.

Bro Tiger Goes Dead

by James Berry

Tiger swears he's going to crack up Anancy's bones once and for all.

Tiger goes to bed. Bro Tiger lies down in his bed, all still and stiff, wrapped up in a sheet. Bro Tiger says to himself, 'I know that Anancy will come and look at me. The brute will want to make sure I'm dead. He'll want to see how I look when I'm dead. That's when I'm going to collar him up. Oh, how I'm going to grab that Anancy and finish him!'

Bro Tiger calls his wife. He tells his wife she should begin to bawl. She should bawl and cry and wail as loud as she can. She should stand in the yard, put her hands on the top of her head and holler to let everybody know her husband is dead. And Mrs Tiger does that.

Mrs Tiger bawls and bawls so loud that people begin to wonder if all her family is dead suddenly and not just her husband.

Village people come and crowd in the yard, quick-quick. Everybody is worried and sad and full of sympathy. The people talk to one another saying, 'Fancy how Bro Tiger is dead, sudden-sudden.'

'Yes! Fancy how he's dead sudden-sudden. All dead and gone!'

Anancy also hears the mournful death howling. When Anancy hears it, listen to the Anancy to himself. 'Funny how Bro Tiger is dead. Bro Tiger is such a strong and healthy man. Bro Tiger is such a well-fed man. Bro Tiger is dead and I've heard nothing about his sickness.'

Anancy finds himself at Tiger's yard, like the rest of the crowd.

Straightaway, Anancy says to his son, 'Tacooma, did you happen to hear Bro Tiger had an illness?'

Tacooma shakes his head. 'No, no. Heard nothing at all.'

Anancy goes to Dog. 'Bro Dog, did you happen to hear Bro Tiger had an illness?'

Bro Dog shakes his head. 'No, no. Heard nothing at all.'

Anancy goes to Monkey and Puss and Ram-Goat and Jackass and Patoo and asks the same question. Everyone gives a sad shake of the head and says, 'No, no. Heard nothing at all.'

The crowd surrounds Anancy. Everybody starts up saying, 'Bro Tiger showed no sign of illness. Death happened so sudden-sudden, Bro Nancy. So sudden-sudden!'

Anancy says, 'Did anybody call a doctor?'

The people shake their heads and say, 'That would have been no use, Bro Nancy. No use at all.'

'Before death came on, did Tiger call the name of the Lord? Did he whimper? Did he cry out?'

'He didn't have time, Bro Nancy. He didn't have the time,' everybody says. 'It was all so sudden.'

Listen to Anancy now, talking at the top of his voice.

'What kind of man is Tiger? Doesn't Tiger know that no good man can meet his Blessed Lord sudden-sudden and not shudder and cry out?'

Tiger hears Anancy. Tiger feels stupid. Tiger feels he has made a silly mistake. Bro Tiger gives the loudest roar he has ever made.

Anancy bursts out laughing. Anancy says, 'Friends, did you hear that? Did you hear that? Has anyone ever heard a dead man cry out?'

Nobody answers Anancy. Everybody sees that he is right.

By the time Bro Tiger jumps out of the sheet on the bed to come after Anancy, the Anancy is gone. Bro Nancy is well away.

Nobody even talks to Bro Tiger now. Everybody just leaves Bro Tiger's place without a single word.

The Wonderful Tar Baby

by George Browne

Now one day Brer Fox is walking along by himself, when he comes on some sticky black tar by the edge of the road. He thinks the road-menders must have left it.

Brer Fox looks at the tar and says to himself, 'This is just the stuff to trap that cheeky Brer Rabbit! I'll make a doll contraption out of some branches and cover it in the sticky tar, and leave it sitting in the road just like a real baby. Brer Rabbit is so sure of himself that when he sees it he's bound to get himself stuck to it. And this time I'll catch that cheeky Brer Rabbit for sure.'

Brer Fox makes the tar baby and puts it in the middle of the big road. Then he goes and hides himself in the bushes to wait and see what will happen. And he doesn't have to wait long, because by and by Brer Rabbit comes pacing down the road — lippety-clippety, clippety-lippety, just as sassy as a jay-bird. When Brer Rabbit sees the tar baby he's astonished.

'Morning!' says Brer Rabbit says he. 'Nice weather this morning.'

But the tar baby ain't saying nothing. It can't of course, because it's only made of tar. Behind the bushes, Brer Fox lays low and winks his eye slow.

'How is you this morning?' says Brer Rabbit to the tar baby. 'Is you deaf? Because if you is, I can holler louder,' says he. But the tar

baby stays still and Brer Fox, he lays low and chuckles in his stomach.

'You is just stuck up, that's what you is,' says Brer Rabbit to the tar baby. 'And I'm going to learn you to talk to respectable folk if it's the last thing I do,' says Brer Rabbit, getting very angry.

'If you don't say "How do you do," I'm going to bust you wide open.' But the tar baby keeps on saying nothing. Then Brer Rabbit draws back his fist and hits the tar baby, and his fist sticks in the tar, and he can't pull it free — the stickiness holds him fast.

'Turn me loose or I'll bash you,' says Brer Rabbit, and with that he hits the tar baby with his other fist, and that's stuck fast as well. But the tar baby ain't saying nothing. Then Brer Rabbit hits the tar baby with his feet, and butts it with his head — until he's stuck fast all over.

Then Brer Fox saunters out from behind the bushes, looking just as innocent as a mocking-bird.

'Howdy, Brer Rabbit!' says Brer Fox. 'You look sort of stuck up this morning!' And he laughs and he laughs till he can't laugh no more.

'Well I expect I've got you this time,' says Brer Fox. 'Maybe I ain't but I expect I have. You've been running around here all this time, bouncing round the neighbourhood until you've come to believe yourself the boss of the whole gang. Who asked you for to try to strike up an acquaintance with this here tar baby? And who stuck you up where you are? Nobody in the whole round world! You just stuck yourself on that tar, Brer Rabbit, without waiting for anyone to invite you,' says Brer Fox, 'and there you are and there'll you stay till I fixes up a fire. I'm going to cook you for my dinner, Brer Rabbit, this day, that's for certain.'

Then Brer Rabbit has to pretend to be mighty humble.

'Brer Fox,' says he, 'fix up a fire and cook me if you like — but whatever you do, please don't fling me into the thorns on that briar patch!'

'What's that?' says Brer Fox.

'Fix up a fire and cook me if you like,' says Brer Rabbit. 'I don't care what you do with me, just so long as you don't fling me in that briar patch.'

That makes Brer Fox stop to think. 'I don't know whether I'll cook you,' says Brer Fox, 'it's too much trouble to kindle a fire. I expect I'll hang you from that tree.'

'Hang me as high as you please, Brer Fox, but for the Lord's sake, don't fling me in that briar patch.'

'I ain't got no string to hang you with,' says Brer Fox, 'so now I expect I'll have to drown you.'

'Drown me as deep as you please, Brer Fox, but please, please, don't fling me in that briar patch.'

Now Brer Fox wants to hurt Brer Rabbit just as bad as he can, and he thinks that the thorns in the briar patch will scratch him to pieces. So he catches him by the hind legs and flings him right out into the middle of the briar patch.

There's quite a flutter where Brer Rabbit hits the thorn bushes, and Brer Fox sort of hangs around to see what's going to happen next.

Then by and by he hears someone calling him from way out up on the hill and there's Brer Rabbit singing out, mighty pleased with himself.

'Thorns can't hurt me, Brer Fox!
Born and bred in the briar patch,
Brer Fox!
Lived all my life in the briar patch,
Brer Fox,
Born and bred in the briar patch,
Brer Fox.'

And off he slips as lively as a cricket. So Brer Rabbit gets safe away from Brer Fox for another day.

Simple Peter's Mirror

by Terry Jones

Simple Peter was walking to work in the fields one morning when he met an old woman sitting beside the road.

'Good morning, old woman,' he said, 'why do you look so sad?'

'I have lost my ring,' said the old woman, 'and it is the only one like it in the whole world.'

'I will help you find it,' said Simple Peter, and he got down on his hands and knees to look for the old woman's ring.

Well, he hunted for a long time, until at last he found the ring under a leaf.

'Thank you,' said the old woman. 'That ring is more precious than you realize,' and she slipped it on to her finger. Then she took a mirror out of her apron and gave it to Peter, saying, 'Take this as a reward.'

Now Simple Peter had never seen a mirror before and so, when he looked down and saw the reflection of the sky in his hands, he said, 'Have you given me the sky?'

'No,' said the old woman, and explained what it was.

'What do I want with a mirror?' asked Peter.

'That is no ordinary mirror,' replied the old woman. 'It is a magic mirror. Anyone who looks into it will see themselves not as they are, but as other people see them. And that's a great gift, you know, to see ourselves as others see us.'

Simple Peter held the mirror up to his face and peered into it. First he turned one way, then he turned the other. He held the mirror sideways and longways and upside down, and finally he shook his head and said, 'Well, it may be a magic mirror, but it's no good to me, I can't see myself in it at all.'

The old woman smiled and said, 'The mirror will never lie to you. It will show you a true reflection of yourself as other people see you.' And with that she touched her ring, and the oak tree that was standing behind her bent down and picked her up in its branches, and carried her away.

Well, Simple Peter stood there gaping for a long while, and then he looked in the mirror again, and still he could not see himself, even when he put his nose right up against it.

Just then a farmer came riding past on his way to market. 'Excuse me,' said Simple Peter, 'but have you seen my reflection? I can't find it in this mirror.'

'Oh,' said the farmer, 'I saw it half an hour ago, running down the road.'

'Thank you,' said Peter, 'I'll see if I can catch it,' and he ran off down the road.

The farmer laughed and said to himself, 'That Simple Peter is a proper goose!' and he went on his way.

Simple Peter ran on and on until he came to the blacksmith.

'Where are you running so fast, Peter?' called the blacksmith.

'I'm trying to catch my reflection,' replied Peter. 'John the farmer said it ran this way. Did you see it?'

The blacksmith, who was a kindly man,

shook his head and said, 'John the farmer has been telling you stories. Your reflection can't run away from you. Look in the mirror, and you'll see it there all right.'

So Peter looked in the magic mirror, and do you know what he saw? He saw a goose, with a yellow beak and black eyes, staring straight back at him.

'There, do you see your reflection?' asked the blacksmith.

'I only see a goose,' said Peter, 'but *I'm* not a goose. I'll show you all! I'll seek my fortune, and then you'll see me as I really am!'

So Peter set off to seek his fortune.

Before long, he came to a wild place in the mountains, where he met a woodcutter and his family with all their belongings on their backs.

'Where are you going?' he asked them.

'We're leaving this country,' said the woodcutter, 'because there is a dragon here. It is fifty times as big as a man and could eat you up in one mouthful. Now it has carried off the King's daughter and is going to eat her for supper tonight.' And with that they hurried on their way.

Peter went on, and the mountains grew steeper and the way became harder. All at once he heard a sound like a grindstone. He looked

round a rock and there he saw the dragon. Sure enough, it was fifty times as big as himself and it was spinning a stone round in its front claws to sharpen its teeth.

'Oh ho! Are you the dragon?' asked Peter. The dragon stopped sharpening its teeth and glared with great fierce eyes at Peter.

'I am!' said the dragon.

'Then I shall have to kill you,' said Peter.

'*Indeed?*' said the dragon, and the spines on its back started to bristle, and tongues of flame began to leap out of its nostrils. 'And how are *you* going to kill *me*?'

And Peter said, 'Oh, *I'm* not, but behind this rock I have the most terrible creature, that is fifty times as big as you, and could eat *you* up in one mouthful!'

'Impossible!' roared the dragon, and leapt behind the rock. Now Peter, who was not *so* simple after all, had hidden the magic mirror there, and so when the dragon came leaping round the rock it ran slap bang into it, and there, for the first time, it saw itself as it appeared to others — fifty times as big and able to eat itself up in one mouthful! And then and there that dragon turned on its tail and ran off over the mountains as fast as it could, and was never seen again.

Then Peter went into the dragon's cave, and found the King's daughter, and carried her back to the palace. And the King gave him jewels and fine clothes and all the people cheered him to the skies. And when Peter looked in the magic mirror now, do you know what he saw? He saw himself as a brave, fierce lion, which was how everyone else saw him. But he said to himself, 'I'm not a lion! I'm Peter.'

Just then the Princess came by and Peter showed her the mirror and asked her what she saw there.

'I see the most beautiful girl in the world,' said the Princess. 'But *I'm* not the most beautiful girl in the world.'

'But that's how you appear to me,' said Peter, and he told the Princess the whole story about how he had come by the mirror, and how he had tricked the dragon.

'So you see, I'm not really a goose, and I'm not really as brave as a lion. I'm just Simple Peter.'

When the Princess heard this story, she began to like him for his straightforwardness and honesty. Pretty soon she grew to love him, and the King agreed that they should be married, even though Peter was just a poor ploughman's son.

'But, my dear!' said the Queen. 'People will make fun of us because he is not a real prince.'

'Fiddlesticks!' replied the King. 'We'll make him into the finest prince you ever did see!' But the old Queen was right. . . .

On the day of the wedding, Peter was dressed up in the finest clothes, trimmed with gold and fur. But when he looked in the magic mirror, do you know what he saw? Instead of a rich and magnificent prince, he saw himself in his own rags — Simple Peter. But it didn't worry him. He smiled and said to himself, 'At last! Everyone sees me as I really am!'

The Tinderbox

by Hans Christian Andersen

translated by Erik Haugaard

A soldier came marching down the road: Left . . . right! Left . . . right! He had a pack on his back and a sword at his side. He had been in the war and he was on his way home. Along the road he met a witch. She was a disgusting sight, with a lower lip that hung all the way down to her chest.

'Good evening, young soldier,' she said. 'What a handsome sword you have and what a big knapsack. I can see that you are a real soldier! I shall give you all the money that you want.'

'Thank you, old witch,' he said.

'Do you see that big tree?' asked the witch, and pointed to the one they were standing next to. 'The trunk is hollow. You climb up to the top of the tree, crawl into the hole, and slide deep down inside it. I'll tie a rope around your waist, so I can pull you up again when you call me.'

'What am I supposed to do down in the tree?' asked the soldier.

'Get money!' answered the witch and laughed. 'Now listen to me. When you get down to the very bottom, you'll be in a great passageway where you'll be able to see because there are over a hundred lamps burning. You'll find three doors; and you can open them all because the keys are in the locks. Go into the first one; and there on a chest, in the middle of the room, you'll see a dog with eyes as big as teacups. Don't let that worry you. You will have my blue checked apron; just spread it out on the floor, put the dog down on top of it, and it won't do you any harm. Open the chest and take as many coins as you wish, they are all copper. If it's silver you're after, then go into the next room. There you'll find a dog with eyes as big as millstones; but don't let that worry you, put him on the apron and take the money. If you'd rather have gold, you can have that too; it's in the third room. Wait till you see that dog, he's got eyes as big as the Round Tower in Copenhagen; but don't let that worry you. Put him down on my apron and he won't hurt you; then you can take as much gold as you wish.'

'That doesn't sound bad!' said the soldier. 'But what am I to do for you, old witch? I can't help thinking that you must want something too.'

'No,' replied the witch. 'I don't want one single coin. Just bring me the old tinderbox that my grandmother forgot the last time she was down there.'

'I'm ready, tie the rope around my waist!' ordered the soldier.

'There you are, and here is my blue checked apron,' said the witch.

The soldier climbed the tree, let himself fall into the hole, and found that he was in the passageway, where more than a hundred lights burned.

He opened the first door. Oh! There sat the dog with eyes as big as teacups glaring at him.

'You are a handsome fellow!' he exclaimed as he put the dog down on the witch's apron. He filled his pockets with copper coins, closed the chest, and put the dog back on top of it.

He went into the second room. Aha! There sat the dog with eyes as big as millstones. 'Don't keep looking at me like that,' said the soldier good-naturedly. 'It isn't polite and you'll spoil your eyes.' He put the dog down on the witch's apron and opened the chest. When he saw all the silver coins, he emptied the copper out of his pockets and filled both them and his knapsack with silver.

56

Now he entered the third room. That dog was big enough to frighten anyone, even a soldier. His eyes were as large as the Round Tower in Copenhagen and they turned around like wheels.

'Good evening,' said the soldier politely, taking off his cap, for such a dog he had never seen before. For a while he just stood looking at it; but finally he said to himself, 'Enough of this!' Then he put the dog down on the witch's apron and opened up the chest.

'God preserve me!' he cried. There was so much gold that there was enough to buy the whole city of Copenhagen; and all the gingerbread men, rocking horses, riding whips, and tin soldiers in the whole world.

Quickly the soldier threw away all the silver coins that he had in his pockets and knapsack and put gold in them instead; he even filled his boots and his cap with money. He put the dog back on the chest, closed the door behind him, and called up through the hollow tree.

'Pull me up, you old witch!'

'Have you got the tinderbox?' she called back.

'Right you are, I have forgotten it,' he replied honestly, and went back to get it. The witch hoisted him up and again he stood on the road; but now his pockets, knapsack, cap, and boots were filled with gold and he felt quite differently.

'Why do you want the tinderbox?' he asked.

'Mind your own business,' answered the witch crossly. 'You have got your money, just give me the tinderbox.'

'Blah! Blah!' said the soldier. 'Tell me what you are going to use it for, right now; or I'll draw my sword and cut off your head.'

'No!' replied the witch firmly; but that was a mistake, for the soldier chopped her head off. She lay there dead. The soldier put all his gold in her apron, tied it up into a bundle, and threw it over his shoulder. The tinderbox he dropped into his pocket; and off to town he went.

The town was nice, and the soldier went to the nicest inn, where he asked to be put up in the finest room and ordered all the things he liked to eat best for his supper, because now he had so much money that he was rich.

The servant who polished his boots thought it was very odd that a man so wealthy should have such worn-out boots. But the soldier hadn't had time to buy anything yet; the next day he bought boots and clothes that fitted his purse. And the soldier became a refined gentleman. People were eager to tell him all about their town and their king, and what a lovely princess his daughter was.

'I would like to see her,' said the soldier.

'But no one sees her,' explained the townfolk. 'She lives in a copper castle, surrounded by walls, and towers, and a moat. The king doesn't dare allow anyone to visit her because it has been foretold that she will marry a simple soldier, and the king doesn't want that to happen.'

'If only I could see her,' thought the soldier, though it was unthinkable.

The soldier lived merrily, went to the theatre, kept a carriage so he could drive in the king's park, and gave lots of money to the poor. He remembered well what it felt like not to have a penny in his purse.

He was rich and well dressed. He had many friends; and they all said that he was kind and a real cavalier; and such things he liked to hear. But since he used money every day and never received any, he soon had only two copper coins left.

He had to move out of the beautiful room downstairs, up to a tiny one in the garret, where he not only polished his boots himself but also mended them with a large needle. None of his friends came to see him, for they said there were too many stairs to climb.

It was a very dark evening and he could not even buy a candle. Suddenly he remembered that he had seen the stub of a candle in the tinderbox that he had brought up from the bottom of the hollow tree. He found the tinderbox and took out the candle. He struck the flint. There was a spark, and in through the door came the dog with eyes as big as teacups. 'What does my master command?' asked the dog.

'What's this all about?' exclaimed the soldier. 'That certainly was an interesting tinderbox. Can I have whatever I want? Bring me some money,' he ordered. In less time than it takes to say thank you, the dog was gone and back with a big sack of copper coins in his mouth.

Now the soldier understood why the witch had thought the tinderbox so valuable. If he struck it once, the dog appeared who sat on the chest full of copper coins; if he struck it twice, then the dog came who guarded the silver money; and if he struck it three times, then came the one who had the gold.

The soldier moved downstairs again, wore fine clothes again, and had fine friends, for now they all remembered him and cared for him as they had before.

One night, when he was sitting alone after his friends had gone, he thought, 'It is a pity that no one can see that beautiful princess. What is the good of her beauty if she must always remain behind the high walls and towers of a copper castle? Will I never see her? . . . Where is my tinderbox?'

He made the sparks fly and the dog with eyes as big as teacups came. 'I know it's very late at night,' he said, 'but I would so like to see the beautiful princess, if only for a minute.'

Away went the dog; and faster than thought he returned with the sleeping princess on his back. She was so lovely that anyone would have known that she was a real princess. The soldier could not help kissing her, for he was a true soldier.

The dog brought the princess back to her copper castle; but in the morning while she was having tea with her father and mother, the king and queen, she told them that she had had a very strange dream that night. A large dog had come and carried her away to a soldier who kissed her.

'That's a nice story,' said the queen, but she didn't mean it.

The next night one of the older ladies-in-waiting was sent to watch over the princess while she slept, and find out whether it had only been a dream, and not something worse.

The soldier longed to see the princess so much that he couldn't bear it, so at night he sent the dog to fetch her.

The dog ran as fast as he could, but the lady-in-waiting had her boots on and she kept up with him all the way. When she saw which house he had entered, she took out a piece of chalk and made a big white cross on the door.

'Now we'll be able to find it in the morning,' she thought, and went home to get some sleep.

When the dog returned the princess to the castle, he noticed the cross on the door of the house where his master lived; so he took a piece of white chalk and put crosses on all the doors of all the houses in the whole town. It was a very clever thing to do, for now the lady-in-waiting would never know which was the right door.

The next morning the king and queen, the old lady-in-waiting, and all the royal officers went out into town to find the house where the princess had been.

'Here it is!' exclaimed the king, when he saw the first door with a cross on it.

'No, my sweet husband, it is here,' said his wife, who had seen the second door with a cross on it.

'Here's one!'

'There's one!'

Everyone shouted at once, for it didn't matter where anyone looked: there he would find a door with a cross on it; and so they all gave up.

Now the queen was so clever, she could do more than ride in a golden carriage. She took out her golden scissors and cut out a large piece of silk and sewed it into a pretty little bag. This she filled with the fine grain of buckwheat, and tied the bag around the princess's waist. When this was done, she cut a little hole in the bag just big enough for the little grains of buckwheat to fall out, one at a time, and show the way to the house where the princess was taken by the dog.

During the night the dog came to fetch the princess and carry her on his back to the soldier, who loved her so much that now he had only one desire, and that was to be a prince so that he could marry her.

The dog neither saw nor felt the grains of buckwheat that made a little trail all the way from the copper castle to the soldier's room at the inn. In the morning the king and queen had no difficulty in finding where the princess had been, and the soldier was thrown into jail.

There he sat in the dark with nothing to do; and what made matters worse was that everyone said, 'Tomorrow you are going to be hanged!'

That was not amusing to hear. If only he had had his tinderbox, but he had forgotten it in his room. When the sun rose, he watched the people, through the bars of his window, as they hurried towards the gates of the city, for the hanging was to take place outside the walls. He heard the drums and the royal soldiers marching. Everyone was running. He saw a shoemaker's apprentice, who had not bothered to take off his leather apron and was wearing slippers. The boy lifted his legs so high, it looked as though he were galloping. One of his slippers flew off and landed near the window of the soldier's cell.

'Hey!' shouted the soldier. 'Listen, shoemaker, wait a minute, nothing much will happen before I get there. But if you will run to the inn and get the tinderbox I left in my room, you can earn four copper coins. But you'd better use your legs or it will be too late.'

The shoemaker's apprentice, who didn't have one copper coin, was eager to earn four; and he ran to get the tinderbox as fast as he could and gave it to the soldier.

And now you shall hear what happened after that!

Outside the gates of the town, a gallows had been built; around it stood the royal soldiers and many hundreds of thousands of people. The king and the queen sat on their lovely throne, and opposite them sat the judge and the royal council.

The soldier was standing on the platform, but as the noose was put around his neck, he declared that it was an ancient custom to grant a condemned man his last innocent wish. The only thing he wanted was to be allowed to smoke a pipe of tobacco.

The king couldn't refuse, and the soldier took out his tinderbox and struck it: once, twice, three times! Instantly, the three dogs were before him: the one with eyes as big as teacups, the one with eyes as big as millstones, and the one with eyes as big as the Round Tower in Copenhagen.

'Help me! I don't want to be hanged!' cried the soldier.

The dogs ran towards the judge and the royal council. They took one man by the leg and another by the nose, and threw them up in the air, so high that when they hit the earth again they broke into little pieces.

'Not me!' screamed the king, but the biggest dog took both the king and the queen and sent them flying up as high as all the others had been.

The royal guards got frightened, and the people began to shout: 'Little soldier, you shall be our king and marry the princess!'

The soldier rode in the king's golden carriage, and the three dogs danced in front of it and barked: 'Hurrah!'

The little boys whistled and the royal guards presented arms. The princess came out of her copper castle and became queen, which she liked very much. The wedding feast lasted a week, and the three dogs sat at the table and made eyes at everyone.

The Dauntless Girl
by Kevin Crossley-Holland

'Dang it!' said the farmer.

'Why?' said the miller.

'Not a drop left,' the farmer said.

'Not one?' asked the blacksmith, raising his glass and inspecting it. His last inch of whisky glowed like molten honey in the flickering firelight.

'Why not?' said the miller.

'You fool!' said the farmer. 'Because the bottle's empty.' He peered into the flames. 'Never mind that though,' he said. 'We'll send out my Mary. She'll go down to the inn and bring us another bottle.'

'What?' said the blacksmith. 'She'll be afraid to go out on such a dark night, all the way down to the village, and all on her own.'

'Never!' said the farmer. 'She's afraid of nothing — nothing live or dead. She's worth all my lads put together.'

The farmer gave a shout and Mary came out of the kitchen. She stood and she listened. She went out into the dark night and in a little time she returned with another bottle of whisky.

The miller and the blacksmith were delighted. They drank to her health, but later the miller said, 'That's a strange thing, though.'

'What's that?' asked the farmer.

'That she should be so bold, your Mary.'

64

'Bold as brass,' said the blacksmith. 'Out and alone and the night so dark.'

'That's nothing at all,' said the farmer. 'She'd go anywhere, day or night. She's afraid of nothing — nothing live or dead.'

'Words,' said the blacksmith. 'But my, this whisky tastes good.'

'Words nothing,' said the farmer. 'I bet you a golden guinea that neither of you can name anything that girl will not do.'

The miller scratched his head and the blacksmith peered at the golden guinea of whisky in his glass. 'All right,' said the blacksmith. 'Let's meet here again at the same time next week. Then I'll name something Mary will not do.'

Next week the blacksmith went to see the priest and borrowed the key of the church door from him. Then he paid a visit to the sexton and showed him the key.

'What do you want with that?' asked the sexton.

'What I want with you,' said the blacksmith, 'is this. I want you to go into the church tonight, just before midnight, and hide yourself in the dead house.'

'Never,' said the sexton.

'Not for half a guinea?' asked the blacksmith.

The old sexton's eyes popped out of his head. 'Dang it!' he said. 'What's that for then?'

'To frighten that brazen farm girl, Mary,' said the blacksmith, grinning. 'When she comes to the dead house, just give a moan or a holler.'

The old sexton's desire for the half guinea was even greater than his fear. He hummed and hawed and at last he agreed to do as the blacksmith asked.

Then the blacksmith clumped the sexton on the back with his massive fist and the old sexton coughed.

'I'll see you tomorrow,' said the blacksmith, 'and settle the account.

Just before midnight, then! Not a minute later!'

The sexton nodded and the blacksmith strode up to the farm. Darkness was falling and the farmer and the miller were already drinking and waiting for him.

'Well?' said the farmer.

The blacksmith grasped his glass then raised it and rolled the whisky around his mouth.

'Well,' said the farmer. 'Are you or aren't you?'

'This,' said the blacksmith, 'is what your Mary will not do. She won't go into the church alone at midnight . . .'

'No,' said the miller.

'. . . and go to the dead house,' continued the blacksmith, 'and bring back a skull bone. That's what she won't do.'

'Never,' said the miller.

The farmer gave a shout and Mary came out of the kitchen. She stood and she listened; and later, at midnight, she went out into the darkness and walked down to the church.

Mary opened the church door. She held up her lamp and clattered down the steps to the dead house. She pushed open its creaking door and saw skulls and thigh bones and bones of every kind gleaming in front of her. She stooped and picked up the nearest skull bone.

'Let that be,' moaned a muffled voice from behind the dead house door. 'That's my mother's skull bone.'

So Mary put that skull down and picked up another.

'Let that be,' moaned a muffled voice from behind the dead house door. 'That's my father's skull bone.'

So Mary put that skull bone down too and picked up yet another one. And, as she did so, she angrily called out, 'Father or mother, sister or brother, I *must* have a skull bone and that's my last word.' Then she walked out of the dead house, slammed the door, and hurried up the steps and back up to the farm.

Mary put the skull bone on the table in front of the farmer.

'There's your skull bone, master,' she said, and started off for the kitchen.

'Wait a minute!' said the farmer, grinning and shivering at one and the same time. 'Didn't you hear anything in the dead house, Mary?'

'Yes,' she said. 'Some fool of a ghost called out to me: "Let that be! That's my mother's skull bone" and "Let that be! That's my father's skull bone." But I told him straight: "Father or mother, sister or brother, I *must* have a skull bone".'

The miller and the blacksmith stared at Mary and shook their heads.

'So I took one,' said Mary, 'and here I am and here it is.' She looked down at the three faces flickering in the firelight. 'As I was going away,' she said, 'after I had locked the door, I heard the old ghost hollering and shrieking like mad.'

The blacksmith and the miller looked at each other and got to their feet.

'That'll do then, Mary,' said the farmer.

The blacksmith knew that the sexton must have been scared out of his wits at being locked all alone in the dead house. They all raced down to the church, and clattered down the steps into the dead house, but they were too late. They found the old sexton lying stone dead on his face.

'That's what comes of trying to frighten a poor young girl,' said the farmer.

So the blacksmith gave the farmer a golden guinea and the farmer gave it to his Mary.

Mary and her daring were known in every house. And after her visit to the dead house, and the death of the old sexton, her fame spread for miles and miles around.

One day the squire, who lived three villages off, rode up to the farm and asked the farmer if he could talk to Mary.

'I've heard,' said the squire, 'that you're afraid of nothing.'

Mary nodded.

'Nothing live or dead,' said the farmer proudly.

'Listen then!' said the squire. 'Last year my old mother died and was buried. But she will not rest. She keeps coming back into the house, and especially at mealtimes.'

Mary stood and listened.

'Sometimes you can see her, sometimes you can't. And when you can't, you can still see a knife and fork get up off the table and play about where her hands would be.'

'That's a strange thing altogether,' said the farmer, 'that she should go on walking.'

'Strange and unnatural,' said the squire. 'And now my servants won't stay with me, not one of them. They're all afraid of her.'

The farmer sighed and shook his head. 'Hard to come by, good servants,' he said.

'So,' said the squire, 'seeing as she's afraid of nothing, nothing live or dead, I'd like to ask your girl to come and work with me.'

Mary was pleased at the prospect of such good employment and, sorry as he was to lose her, the farmer saw there was nothing for it but to let her go.

'I'll come,' said the girl. 'I'm not afraid of ghosts. But you ought to take account of that in my wages.'

'I will,' said the squire.

So Mary went back with the squire to be his servant. The first thing she always did was to lay a place for the ghost at table, and she took great care not to let the knife and fork lie criss-cross.

At meals, Mary passed the ghost the meat and vegetables and sauce and gravy. And then she said: 'Pepper, madam?' and 'Salt, madam?'

The ghost of the squire's mother was pleased enough. Things went on the same from day to day until the squire had to go up to London to settle some legal business.

Next morning Mary was down on her knees, cleaning the parlour grate, when she noticed something thin and glimmering push in through the parlour door, which was just ajar; when it got inside the room, the shape began to swell and open out. It was the old ghost.

For the first time, the ghost spoke to the girl. 'Mary,' she said in a hollow voice, 'are you afraid of me?'

'No, madam,' said Mary, 'I've no cause to be afraid of you, for you are dead and I'm alive.'

For a while the ghost looked at the girl kneeling by the parlour grate. 'Mary,' she said, 'will you come down into the cellar with me? You mustn't bring a light — but I'll shine enough to light the way for you.'

So the two of them went down the cellar steps and the ghost shone like an old lantern. When they got down to the bottom, they went down a passage, and took a right turn and a left, and then the ghost pointed to some loose tiles in one corner. 'Pick up those tiles,' she said.

Mary did as she was asked. And underneath the tiles were two bags of gold, a big one and a little one.

The ghost quivered. 'Mary,' she said, 'that big bag is for your master. But that little bag is for you, for you are a dauntless girl and deserve it.'

Before Mary could open the bag or even open her mouth, the old ghost drifted up the steps and out of sight. She was never seen again and Mary had a devil of a time groping her way along the dark passage and up out of the cellar.

After three days, the squire came back from London.

'Good morning, Mary,' he said. 'Have you seen anything of my mother while I've been away?'

'Yes, sir,' said Mary. 'That I have.' She opened her eyes wide. 'And if you aren't afraid of coming down into the cellar with me, I'll show you something.'

The squire laughed. 'I'm not afraid if you're not afraid,' he said, for the dauntless girl was a very pretty girl.

So Mary lit a candle and led the squire down into the cellar, walked down the passage, took a right turn and a left, and raised the loose tiles in the corner for a second time.

'Two bags,' said the squire.

'Two bags of gold,' said Mary. 'The little one is for you and the big one is for me.'

'Lor!' said the squire, and he said nothing else. He did think that his mother might have given him the big bag, as indeed she had, but all the same he took what he could.

After that, Mary always crossed the knives and forks at meal-times to prevent the old ghost from telling what she had done.

The squire thought things over: the gold and the ghost and Mary's good looks. What with one thing and another he proposed to Mary, and the dauntless girl, she accepted him. In a little while they married, and so the squire did get both bags of gold after all.

The Riddlemaster

by Catherine Storr

Sitting on one of the public benches in the High Street one warm Saturday morning, Polly licked all round the top of an ice-cream horn.

A large person sat down suddenly beside her. The bench swayed and creaked, and Polly looked round.

'Good morning, Wolf!'

'Good morning, Polly.'

'Nice day, Wolf.'

'Going to be hot, Polly.'

'Mmm,' Polly said. She was engaged in trying to save a useful bit of ice-cream with her tongue before it dripped on to the pavement and was wasted.

'In fact it is hot now, Polly.'

'I'm not too hot,' Polly said.

'Perhaps that delicious looking ice is cooling you down,' the wolf said enviously.

'Perhaps it is,' Polly agreed.

'I'm absolutely boiling,' the wolf said.

Polly fished in the pocket of her cotton dress and pulled out a

threepenny bit. It was more than half what
she had left, but she was a kind girl, and
in a way she was fond of the wolf,
tiresome as he sometimes was.

'Here you are, Wolf,' she said,
holding it out to him. 'Go into
Woolworths and get one for yourself.'

There was a scurry of feet, a flash of
black fur, and a little cloud of white
summer dust rose off the pavement near
Polly's feet. The wolf had gone.

Two minutes later he came back, a good deal
more slowly. He was licking his ice-cream horn with a very long red
tongue and it was disappearing extremely quickly. He sat down again
beside Polly with a satisfied grunt.

'Mm! Just what I needed. Thank you very much, Polly.'

'Not at all, Wolf,' said Polly, who had thought that he might have
said this before.

She went on licking her ice in a happy dream-like state, while the
wolf did the same, but twice as fast.

Presently, in a slightly aggrieved voice, the wolf said, 'Haven't you
nearly finished?'

'Well no, not nearly,' Polly said. She always enjoyed spinning out
ices as long as possible. 'Have you?'

'Ages ago.'

'I wish you wouldn't look at me so hard, Wolf,' Polly said,
wriggling. 'It makes me feel uncomfortable when I'm eating.'

'I was only thinking,' the wolf said.

'You look sad, then, when you think,' Polly remarked.

'I generally am. It's a very sad world, Polly.'

'Is it?' said Polly, in surprise.

'Yes. A lot of sad things happen.'

'What things?' asked Polly.

'Well, I finish up all my ice-cream.'

'That's fairly sad. But at any rate you did have it,' Polly said.

'I haven't got it Now,' the wolf said. 'And it's Now that I want it. Now is the only time to eat ice-cream.'

'When you are eating it, it is Now,' Polly remarked.

'But when I'm not, it isn't. I wish it was always Now,' the wolf sighed.

'It sounds like a riddle,' Polly said.

'What does?'

'What you were saying. When is Now not Now or something like that. You know the sort I mean, when is a door not a door?'

'I love riddles,' said the wolf in a much more cheerful voice. 'I know lots. Let's ask each other riddles.'

'Yes, let's,' said Polly.

'And I tell you what would make it really amusing. Let's say that whoever wins can eat the other person up.'

'Wins how?' Polly asked, cautiously.

'By asking three riddles the other person can't answer.'

'Three in a row,' Polly insisted.

'Very well. Three in a row.'

'And I can stop whenever I want to.'

'All right,' the wolf agreed, unwillingly. 'And I'll start,' he added quickly. 'What made the penny stamp?'

Polly knew it was because the threepenny bit, and said so. Then she asked the wolf what made the apple turnover, and he knew the answer to that. Polly knew what was the longest word in the dictionary, and the wolf knew what has an eye but cannot see. This reminded him of the question of what has hands, but no fingers and a face, but no nose, to which Polly was able to reply that it was a clock.

'My turn,' she said, with relief.

'Wolf, what gets bigger, the more you take away from it?'

The wolf looked puzzled.

'Are you sure you've got it right, Polly?' he asked at length. 'You don't mean it gets smaller the more you take away from it?'

'No, I don't.'

'It gets bigger?'

'Yes.'

'No cake I ever saw did that,' the wolf said, thinking aloud. 'Some special kind of pudding, perhaps?'

'It's not a pudding,' Polly said.

'I know!' the wolf said triumphantly. 'It's the sort of pain you get when you're hungry. And the more you don't eat the worse the pain gets. That's getting bigger the less you do about it.'

'No, you're wrong,' Polly said. 'It isn't a pain or anything to eat, either. It's a hole. The more you take away, the bigger it gets, don't you see, Wolf?'

'Being hungry is a sort of hole in your inside,' the wolf said. 'But anyhow it's my turn now. I'm going to ask you a new riddle, so you won't know the answer already, and I don't suppose you'll be able to guess it, either. What gets filled up three or four times a day, and yet can always hold more?'

'Do you mean it can hold more after it's been filled?' Polly asked.

The wolf thought, and then said, 'Yes.'

'But it couldn't, Wolf! If it was really properly filled up it couldn't hold any more.'

'It does though,' the wolf said triumphantly. 'It seems to be quite bursting full and then you try very hard and it still holds a little more.'

Polly had her suspicions of what this might be, but she didn't want to say in case she was wrong.

'I can't guess.'

'It's me!' the wolf cried, in delight. 'Got you that time, Polly! However full up I am, I can always manage a little bit more. Your turn next, Polly.'

'What,' Polly said, 'is the difference between an elephant and a pillar-box?'

The wolf thought for some time.

'The elephant is bigger,' he said, at last.

'Yes. But that isn't the right answer.'

'The pillar-box is red. Bright red. And the elephant isn't.'

'Ye-es. But that isn't the right answer either.'

The wolf looked puzzled. He stared hard at the old-fashioned Victorian pillar-box in the High Street. It had a crimped lid with a knob on top like a silver teapot. But it didn't help him. After some time he said crossly, 'I don't know.'

'You mean you can't tell the difference between an elephant and a pillar-box?'

'No.'

'Then I shan't send you to post my letters,' Polly said, triumphantly. She thought this was a very funny riddle.

The wolf, however, didn't.

'You don't see the joke, Wolf?' Polly asked, a little disappointed that he was so unmoved.

'I see it, yes. But I don't think it's funny. It's not a proper riddle at all. It's just silly.'

'Now you ask me something,' Polly suggested. After a minute or two's thought, the wolf said, 'What is the difference between pea soup and a clean pocket handkerchief?'

'Pea soup is hot and a pocket handkerchief is cold,' said Polly.

'No. Anyhow you could have cold pea soup.'

'Pea soup is green,' said Polly.

'I expect a clean pocket handkerchief could be green too, if it tried,' said the wolf. 'Do you give it up?'

'Well,' said Polly, 'of course I do know the difference, but I don't know what you want me to say.'

'I want you to say you don't know the difference between them,' said the wolf, crossly.

'But I do,' said Polly.

'But then I can't say what I was going to say!' the wolf cried.

He looked so much disappointed that Polly relented.

'All right, then, you say it.'

'You don't know the difference between pea soup and a clean pocket handkerchief?'

'I'll pretend I don't. No, then,' said Polly.

'You ought to be more careful what you keep in your pockets,' the wolf said. He laughed so much at this that he choked, and Polly had to beat him hard on the back before he recovered and could sit back comfortably on the seat again.

'Your turn,' he said, as soon as he could speak.

Polly thought carefully. She thought of a riddle about a man going to St Ives; of one about the man who showed a portrait to another man; of one about a candle; but she was not satisfied with any of them. With so many riddles it isn't really so much a question of guessing the answers, as of knowing them or not knowing them already, and if the wolf were to invent a completely new riddle out of his head, he would be able to eat her, Polly, in no time at all.

'Hurry up,' said the wolf.

Perhaps it was seeing his long red tongue at such very close quarters, or it may have been the feeling that she had no time to lose, that made Polly say, before she had considered what she was going to say, 'What is it that has teeth, but no mouth?'

'Grrrr,' said the wolf, showing all his teeth for a moment. 'Are you quite sure he hasn't a mouth, Polly?'

'Quite sure. And I'm supposed to be asking the questions, not you, Wolf.'

The wolf did not appear to hear this. He had now turned his back on Polly and was going through some sort of rapid repetition in a subdued gabble, through which Polly could hear only occasional words.

'. . .Grandma, so I said the better to see you with, gabble, gabble, gabble, Ears you've got, gabble gabble better to hear gabble gabble gabble gabble gabble TEETH gabble eat you all up.'

He turned round with a satisfied air.

'I've guessed it, Polly. It's a GRANDMOTHER.'

'No,' said Polly astonished.

'Well then, Red Riding Hood's grandmother if you are so particular. The story mentions her eyes and her ears and her teeth, so I expect she hadn't got anything else. No mouth anyhow.'

'It's not anyone's grandmother.'

'Not a grandmother,' said the wolf slowly. He shook his head. 'It's difficult. Tell me some more about it. Are they sharp teeth, Polly?'

'They can be,' Polly said.

'As sharp as mine?' asked the wolf, showing his for comparison.

'No,' said Polly, drawing back a little. 'But more tidily arranged,' she added.

The wolf shut his jaws with a snap.

'I give up,' he said, in a disagreeable tone. 'There isn't anything I know of that has teeth and no mouth. What use would the teeth be to anyone without a mouth? I mean, what is the point of taking a nice juicy bite out of something if you've got to find someone else's mouth to swallow it for you? It doesn't make sense.'

'It's a comb,' said Polly, when she got a chance to speak.

'A what?' cried the wolf in disgust.

'A COMB. What you do your hair with. It's got teeth, hasn't it? But no mouth. A comb, Wolf.'

The wolf looked sulky. Then he said in a bright voice, 'My turn now, and I'll begin straight away. What is the difference between a nice fat young pink pig and a plate of sausages and bacon? You don't know, of course, so I'll tell you. It's —'

'Wolf!' Polly interrupted.

'It's a very good riddle, this one, and I can't blame you for not having guessed it. The answer is —'

'WOLF!' Polly said, 'I want to tell you something.'

'Not the answer?'

'No. Not the answer. Something else.'

'Well, go on.'

'Look, Wolf, we made a bargain, didn't we, that whoever lost three lives running by not being able to answer riddles, might be eaten up by the other person?'

'Yes,' the wolf agreed. 'And you've lost two already, and now you're not going to be able to answer the third and then I shall eat you up. Now I'll tell you what the difference is between a nice fat little pink —'

'NO!' Polly shouted. 'Listen, Wolf! I may have lost two lives already, but you have lost three!'

'I haven't!'

'Yes, you have! You couldn't answer the riddle about the hole, you didn't know the difference between an elephant and a pillar-box —'

'I do!' said the wolf indignantly.

'Well, you may now, but you didn't when I asked you the riddle; and you didn't know about the comb having teeth and no mouth. That was three you couldn't answer in a row, so it isn't you that is going to eat me up.'

'What is it then?' the wolf asked, shaken.

'It's me that is going to eat you up!' said Polly.

The wolf moved rather further away.

'Are you really going to eat me up, Polly?'

'In a moment, Wolf. I'm just considering how I'll have you cooked,' said Polly.

'I'm very tough, Polly.'

'That's all right, Wolf. I can simmer you gently over a low flame until you are tender.'

'I don't suppose I'd fit very nicely into any of your saucepans, Polly.'

'I can use the big one Mother has for making jam. That's an enormous saucepan,' said Polly, thoughtfully, measuring the wolf with her eyes.

The wolf began visibly to shake where he sat.

'Oh please, Polly, don't eat me. Don't eat me up this time,' he urged. 'Let me off this once, I promise I'll never do it again.'

'Never do what again?' Polly asked.

'I don't know. What was I doing?' the wolf asked himself, in despair.

'Trying to get me to eat,' Polly suggested.

'Well, of course, I'm always doing that,' the wolf agreed.

'And you would have eaten me?' Polly asked.

'Not if you'd asked very nicely, I wouldn't,' the wolf said. 'Like I'm asking now.'

'And if I didn't eat you up, you'd stop trying to get me?'

The wolf considered.

'Look,' he said, 'I can't say I'll stop for ever, because after all a wolf is a wolf, and if I promised to stop for ever I wouldn't be a wolf any more. But I promise to stop for a long time. I won't try any more today.'

'And what about after today?' Polly insisted.

'The first time I catch you,' the wolf said dreamily, 'if you ask *very* nicely I'll let you go because you've let me off today. But after that, no mercy! It'll be just Snap! Crunch! Swallow!'

'All right,' Polly said, recollecting that so far the wolf had not ever got as far as catching her successfully even once. 'You can go.'

The wolf ducked his head gratefully and trotted off. Polly saw him threading his way between the busy shoppers in the High Street.

But she sat contentedly in the hot sun and wondered what was the difference between a fat pink pig and a plate of sausages and bacon. Not much, if she knew her wolf!

Think of a Word

by Joan Aiken

Once there was a boy called Dan who was in the habit of using short rude words.

Almost any short word ending in T was rude in the country where Dan lived: Dit, Fot, Het, Rit, Sut.

'You silly old Sut,' he called after an old lady in the street one day, and she turned round on him, quick as a whiplash.

'You'll be sorry you said that to me,' she said.

'Why, you old Jot?' said Dan.

'Because, from now on,' said the old lady, 'every time you say one of those words you seem so keen on, a square inch of your skin will turn to glass, so that everybody will be able to see all the works inside you. There are eight words that would cure the habit you have,' she said, 'but I shan't teach them to you. You'll have to find them out for yourself.'

And she turned on her skinny old heel and walked away.

Dan was left standing there with his mouth open.

He didn't call anything after the old lady — somehow she had left him rather quiet and thoughtful — but, later in the day, he forgot all about her, and called the driver of the school bus a stupid Nat.

'Coo! Dan!' said his friend Rod who was sitting beside him. 'Your face has gone all funny! I can see your teeth through your cheek as if it was glass. *And* the buttermint you're sucking. You didn't tell *me* you had any buttermints.'

Dan, quite upset, couldn't wait to get home and look in the mirror.

Sure enough, a patch of his right cheek had gone clear and see-through — there were his teeth and his tongue, plain to view.

It was like having a plastic porthole in his face.

And, after two or three days, a good few more patches had gone transparent all over Dan — on his arms, his legs, his neck, and even more inconvenient places. You could see bones and muscles in him, and tubes and joints and things that aren't usually seen.

The family doctor was quite keen to send Dan up to a big teaching hospital, so that the medical students could look at him and find out useful facts. But Dan's mother wasn't having any of that.

She was very annoyed about it, and so was Dan's father.

'It's disgraceful,' they said.

So, since Dan couldn't seem to stop coming out with short rude words ending in T, they took him away from school and sent him off into the mountains to be a shepherd.

Up high in the hills, alone all day with the sheep, he couldn't come to much harm, they reckoned, as there was no one to talk to, and so he wouldn't be using any language, and, by and by, might learn to think before he spoke.

So off went Dan, into the high meadows, where he had no company but the baaing sheep and a surly old dog called Buff, who never barked, and who made it plain to Dan that he could have looked after the whole flock perfectly well on his own, without any help.

There, sitting on a rock, or on the short, sweet mountain grass, Dan had plenty of time to think, and to wonder which were the eight words the old lady had meant.

Day after day he thought, week after week, and he never spoke.

Thoughts piled up inside his head like leaves in a hollow tree. He thought about how you could tow away the wind, if you had a strong enough rope. He thought about how, if you laid your plans carefully, you could win summer or winter to be your very own. He thought about rolled and stuffed thunder, and pan-fried lightning. He thought about weaving a rope of rain. He thought about the air, which is everywhere. He thought about the earth, which is nothing but a shepherd's pie of everything left over.

Words are stronger than blows, he thought. And perhaps, he thought later on, thoughts are stronger than words.

So Dan passed days and weeks and months, wandering among the hills with his sheep. He was happy now. He didn't even want to go back to his home.

He listened to what the wind had to say, he watched the dark and the light playing hide-and-seek with each other, he felt the rock under his toes, he tasted the rain and smelt the warm salt wool of the sheep.

Meanwhile, down in the plains, and in Dan's home town, they were having a lot of trouble with dragons.

Dragons had suddenly started breeding quicker than wasps, and the whole country was full of them. Put your Sunday joint in the oven, and half an hour later a dozen dragons would have smelt it out; they'd be battering at your window like bulldozers.

Dragons fouled up the airport runways with claw-marks and scattered scales and droppings; they burst into banks and snatched bags of cash; they came snorting into cinemas and burned up reels of film; they broke off TV aerials and scraped tiles from roofs; splashing

in rivers they turned all the water to steam; they swallowed down hundreds of men, women, and children going about their daily affairs. And as for princesses — there wasn't a single princess left free in the world, for the dragons had collected the lot, and had them all shut up together in a nasty, greasy, cindery castle, which stood on an island in the middle of a lake, up among the highest peaks of the mountains, which in that part were so tall and sharp that they looked like the spikes of a king's crown.

Dan knew nothing of all this.

He did notice, to be sure, that dragons flew overhead much more than they used to: all of a sudden there would be a big spiny shadow across the sun, and the sheep would bleat in fright and huddle together, and old Buff the sheepdog would growl and snake out his head with flattened ears.

Dan noticed, too, that knights and princes and soldiers were quite often to be seen, riding horses or tanks or motorbikes up the highways into the mountains. From his perch on a high crag Dan would see them go up, but he never saw them come down again. Up, up, the tiny figures went, and vanished into the high passes. Maybe they were crossing the mountains to the other side, Dan reckoned. He didn't give them too much thought. Nor did he trouble his head about the distant rumblings and flashes from those high peaks where they went. A bit of bad weather in the mountains was nothing out of the common. The sheep didn't mind it, nor did Dan.

But one day a young fellow in shining armour, a handsome lad with a ruby-hilted sword and a gold crown around his helmet, came riding past the crag where Dan sat with his flock.

'Good day, shepherd!' called the knight. 'Am I going right for the dragons' castle?'

Dan had to work his jaws and his throat and his tongue for quite a few minutes before he was able to answer — so many months had it been since he had spoken last.

'Umph — dragons' castle?' he croaked out at last. 'Dragons' castle? I'm not sure I know of any dragons' castle.'

'Oh, come *on*! You must know of it! Where they have a hundred princesses shut up together inside — and a hundred dragons on the rampage outside. You mean to say you live up here in the mountains and you haven't heard of that?'

'I mind my own business,' croaked Dan.

But when the knight told him that the castle clung like a cork in a bottle to the tip of a steep island in a mountain lake, Dan was able to set the knight on his right way. 'Up the pass, keep left, round a mountain shaped like a muffin, that'll take you there.'

'*You* seem to live here safe enough, shepherd,' said the knight, rather surprised. 'Aren't you afraid for your flock, with so many dragons about?'

'They can't land here. The slopes are too steep,' Dan told him. 'A dragon needs a flat landing strip, or a stretch of water. Or a crag that he can grab hold of. Slopes are too slippery for them.'

All this Dan brought out very slowly. Finding the words was hard work, and tiring, like a walk through deep mud.

'I can see that you know a lot about dragons,' the knight said, looking at Dan with respect. 'I wonder — can you suggest any way to deal with them?'

'Dragons don't trouble me,' mumbled Dan.

'No — but when I meet one of them — what should I do?'

Dan began to wish that the stranger would go away and leave him in peace.

'Oh,' he said quickly — anything to get rid of the fidgety young fellow — 'just write a word on your forehead with the tip of your finger dipped in morning dew. If you do that, then you'll have power over the dragons.'

'Well, fancy, now!' said the knight. 'What word should I use?'

So Dan quickly told him a word, and he set spurs to his horse and shook the reins. But then, pulling back, he turned and called, 'Don't *you* want to come and rescue those hundred princesses?'

Dan shook his head, and the knight galloped away up the pass.

Sitting down again, Dan gazed at his flock, peacefully nibbling and munching. What? Rescue a hundred princesses? Not likely! Just think of the chattering and giggling and gabbling — the very thought of it made his head buzz. But still, he wished good fortune to the young knight. And now he began to feel a trifle anxious and bothered; for the advice he had given was thought up quite hastily on the spur of the moment. The words had come into his head and he had spoken them. But he hadn't the least notion in the world whether the idea would work or not.

Maybe I ought to go after that young fellow and tell him not to try in case it doesn't work, he thought. Only, if I did that, who would keep an eye on my sheep?

Buff opened one eye and gave a bit of a growl.

What's troubling Buff? Dan wondered. Are there more strangers about?

And then he turned round and noticed a skinny old lady perched nearby on a ledge of rock. Quite comfortable, she looked, and as if she had been there a good long time.

'Found out the use of words, have you, then, Dan?' said she

cordially. And Dan answered her right away, as if the answer had been tucked away in a cupboard of his mind, waiting for this moment:

'Trees are swayed by winds, men by words.'

'Right,' said the old lady, nodding her head energetically. 'And now you've learned that, don't you forget it, Danny my boy. But,' she went on inquisitively, 'what was the word you told that young fellow to write on his forehead?'

'That was a word for *him*,' said Dan. 'Not for any other.'

'Right again,' said the old lady, nodding some more. 'Words are like spices. One is better than a hundred. Learned a bit of sense, you have. Remember it, and maybe you'll be some use in the world by and by.' With that she vanished, like a drop of water off a hot plate, and Dan picked himself a blade of grass and stood chewing it thoughtfully, looking at where she had been.

Next morning early, Dan heard a distant sound that was like the chirping and twittering and chattering of a thousand starlings. And, gazing down at the main highway that led out of the mountains, he saw them going past — what seemed an endless procession of princesses, with their fluttering ribbons and laces and kerchiefs, cloaks and trains and petticoats and veils a-blowing in the wind. A whole hundred of them, in twos and threes, jabbering and jostling, singing and laughing and giggling, down the rocky pass.

I'm glad I'm up here, not down there, thought Dan.

But, by and by, he heard the tramping of a horse's hoofs, and here came the young knight in his gold crown, with a princess, very young and pretty, sitting pillion on the saddle behind him. And a droopy dragon following them, at the end of a long cord.

'*It worked!*' shouted the young knight joyfully. 'It really worked! A thousand, thousand thanks! I'm everlastingly grateful to you — and so are all the princesses.'

The one riding behind him smiled down at Dan, very friendly. She

didn't seem to notice the glass patches all over his skin.

'Won't you come down with us to the city?' said the young man. 'I shall be king one day, and I'll make you my prime minister.'

'No, I thank you, your worship,' said Dan. 'I'd sooner stay here. Besides, people might not respect a prime minister with glass patches all over him. But I'm much obliged for the offer. Only tell me,' he went on, full of curiosity, 'what happened?'

'Why! As soon as the dragons saw the word written in dew on my forehead they all curled up and withered away in flakes of ash. All except this one, which I'm taking to the Zoo. I'd say,' the knight told Dan, 'there wasn't a dragon left now between here and the Western Ocean. Which is all due to you. So I thank you again.'

And with that he set spurs to his horse, and started off down the hill, slipping and sliding, with the dragon limping along behind, and the princess waving thanks and blowing kisses to Dan, until they were out of sight.

All the time they were in view, Dan stood gaping after them. Then he slapped his thighs. Then he began to laugh, and he laughed so hard that he fell down, and Buff stared at him in disapproval.

'It worked!' shouted Dan. 'It really worked! Dragons are bound by cords, and men by words.'

He lay laughing up at the sky, with the larks twittering overhead.

Then he thought: 'What word shall I think of next?'

He damped his finger in the morning dew and wrote on his own forehead.

A growl of thunder rumbled above him, and a lance of lightning flashed like a knitting needle out of a black ball of cloud.

'All right, all right,' shouted Dan, waving gaily to the sky. 'Just keep calm up there, will you? We won't have any of that for the moment. One word at a time is enough.'

And he sat himself down on a rock to watch his sheep.

Mrs Cockle's Cat

by Philippa Pearce

Old Mrs Cockle lived at the top of a very tall house in London. Most of the people who knew her were sorry for her, because she had to climb eighty-four stairs before she reached her own front-door; but she did not mind. It is true that all that climbing made the backs of her knees ache, but then there were advantages. Mrs Cockle lived so high that, from her window, she had a view of the sky over the top of the tall house opposite — which was more than most people had. In the mornings she could look out and think, The sky is blue all over — I'll wear my straw bonnet today; or, The sky is white with snow coming — I'll wear my woollen shawl today; or, The sky has clouded right over — I'll take my biggest umbrella. Mrs Cockle had three umbrellas for different weathers, and the biggest of the three was larger than umbrellas are ever made nowadays.

There was another advantage for Mrs Cockle in living at the very top of the house. In the middle of her ceiling there was a trapdoor, and, if she set up her step-ladder underneath it, she could climb up, open the trapdoor, and climb through on to the roof itself. From the roof she could look round over the buildings of London, and see the factory chimneys and church-spires, and, more than anything else, the chimney-pots — more chimney-pots than you could ever have counted — rows upon rows of chimney-pots that seemed to melt away into the smoky distances of London. This was a fine view, but Mrs Cockle, with the backs of her knees already aching from the eighty-four stairs, would never have bothered about the roof but for Peter.

Peter, who was a cat and lived with Mrs Cockle, was very fond of the roof on a sunny day, or sometimes at night when the moon was full. It was one of the three things that Peter Cockle loved most. The other two things were a little fresh fish for his tea and Mrs Cockle's company. Mrs Cockle, in her turn, was very fond of Peter — more fond of him than of anybody else, for she had no relations, and Mr Cockle had died long before.

Old Mrs Cockle and her cat lived together very contentedly. Every fine day, early, Mrs Cockle opened the trapdoor for Peter to climb on to the roof. Then she set out for her day's work, which was selling coloured balloons at the corner of one of the great London streets. There are plenty of people who sell balloons at street corners in the summer-time or at Christmas; Mrs Cockle was the only person in London who could be counted upon to be selling balloons at her corner, all the year round, day in, day out, and whatever the weather. And, as she said, she did just comfortably enough out of it for herself and Peter.

Late one summer the weather had been particularly wet and blowy, without making any difference to Mrs Cockle; but it had made a difference to Peter. In the first place, he had not cared to venture out on to the roof as usual: that meant that he missed his fresh air and exercise, and felt stuffy and cross, as people do. Besides, he really had something to be cross about. The weather was so bad that the fishermen could not put out to sea as often as usual; there was less fish caught and taken to London and so what little there was in the shops was very dear — too dear for Mrs Cockle to buy. Instead of having fresh fish for his tea every day, Peter had to put up with now a saucer of milk, now some dried haddock, now milk

92

again, now half a tin of herrings in tomato sauce, and so on.

Peter Cockle longed and longed for a mouthful of fresh fish. He knew at last that he loved fresh fish more than a breath of air on the roof, and more even than Mrs Cockle's company. In his own mind, he even blamed Mrs Cockle for the lack of fresh fish, although it was hardly her fault. Nowadays, in the evening, when he and Mrs Cockle sat on either side of the fire with the high wind outside rattling at the windows, they were not as cosy as they used to be. Mrs Cockle sat rocking herself, and knitting, and glancing fondly at Peter. But Peter had given up looking back at her at all: he gazed moodily into the flames of the fire, and saw there nothing but the glittering, slithering shapes of Fresh Fish. And while Mrs Cockle was thinking proudly what a handsome cat he was, Peter was thinking deeply of Fresh Fish until his head seemed to swim with them. And when Mrs Cockle dozed off, Peter was kept awake by the remembrance of Fresh Fish, that seemed to be felt now in his stomach, now on his tongue, and now between his paws.

It was not only in the long quiet evenings that Peter Cockle suffered. One night he could not get to sleep at all for thinking of Fresh Fish. It was then that he took his resolution. Next morning, instead of either mewing to be let out on the roof, or staying curled up on his cushion, he was falsely purring round Mrs Cockle's ankles as she prepared to set out. When she went to the door, he was already there, waiting to say goodbye. But when she opened the door he was out of it ahead of her, and, like a black streak, leaping down the eighty-four steps to the street below.

'Peter! Peter!' called Mrs Cockle distractedly, but there was no sign of his coming back. She climbed down the stairs, hoping to find him waiting for her out in the street. Outside, there was not a cat in sight, except the tabby from next door, and he disliked Mrs Cockle and would not show by as much as a quiver of a whisker which way he had seen Peter go.

'Peter! Peter!' called Mrs Cockle, but softly, because she did not want to start the neighbours talking. No cat came.

Mrs Cockle was naturally upset, but there was nothing to be done, so she went off to her street-corner as usual. She said to herself that the wilful cat would certainly be waiting outside the front door when she went home in the evening. She was only worried about what might happen to him in the meantime. Just suppose, Mrs Cockle thought, he tried to sneak a piece of nice fresh fish from somebody's larder, and they set a dog after him? Or worse, supposing he tried to steal something from the fish shop, and a policeman caught him at it?

That evening Mrs Cockle hurried home a little earlier than usual. There was no Peter in the street outside, and the tabby cat seemed to be sneering. She paid no attention, but hurried up the eighty-four stairs, hoping that, in spite of the tabby cat's expression, Peter might be waiting at the top. But at the very top, there was still no Peter, and poor Mrs Cockle suddenly felt, for the first time in her life, that eighty-four stairs had been too much for her. She sat down on the top one and cried.

94

There was no Peter Cockle that day, nor the next, nor the next. Mrs Cockle tried to go on as usual without him — going to her street-corner, selling balloons, coming home to supper, and going to bed. But her day did not seem the same without a cat: she could not eat properly at mealtimes, nor sleep properly at nights. As time went by she grew thinner and thinner from worry and lack of sleep and lack of food. At last, from being the plumpest balloon-seller in London, she became — so many believed — by far the thinnest. She looked as thin and light as an autumn leaf as she hurried to work along the windy pavements.

One morning, many weeks after Peter's disappearance, Mrs Cockle saw from the sky that the day was going to be wet and very blowy. She put on her goloshes to keep her feet dry, took the largest of her three umbrellas to keep the rest of her dry, and set out to work. It was not yet raining, although the clouds were thick and low overhead; but the wind was already blowing so strongly that she took much longer than usual to reach her street-corner. The hour was still very early and there was no one about. Mrs Cockle hooked her umbrella, still unopened, over her arm, and began blowing up her balloons. Each one, when she had blown it up, she tied for the time being to the iron-fencing. When she had blown them all up — and she had brought out more balloons than usual that morning — she untied them, one by one, to hold them all together in her hand, ready for selling.

At the instant that Mrs Cockle grasped all the balloons in her hand, a great gust of wind came round the corner. It tugged hard at the bunch of balloons, and almost lifted her off her feet, so that she thought she was going to topple over. Instead of falling, however, she felt herself rising. Mrs Cockle's extra thinness and lightness, which had come since Peter's disappearance, and the extra number of balloons that morning, and the extra strong wind at that instant, all helped to lift the old woman clean off her feet, and she began floating upwards. She was so taken by surprise that, for the first few feet upwards, she did not think of calling for help; and after that she felt, as she said later, that she hardly wished to call attention to her position.

Luckily she kept a tight hold on her great bunch of balloons, so that there was no danger of a sudden fall. Even the big umbrella still hung safely from the crook of her arm.

Mrs Cockle went steadily up. She floated up past the first-floor windows of the houses, and saw a young man in pyjamas shaving. She floated past the second-floor windows, and saw two little girls making their beds. Past the third-floor windows, where she saw a whole family sitting down to a breakfast of sausages and bacon. Past the fourth-floor windows, where a fat gentleman was doing exercises to keep down his waist measurements. And lastly, past the attics on the fifth floor, where she saw a boy in bed with measles. He was the only person who saw, first of all, the great bunch of red and blue and yellow balloons, rising steadily, and then, dangling underneath them, Mrs Cockle.

She rose into view and then rose out of it. The boy called his mother in to tell her, but she only thought he must be in a very high fever, and never believed a word.

The balloons and Mrs Cockle were now rising well above the highest of the houses — higher than she and Peter had ever been on their own roof-top. At first a wonderful view of all London appeared beneath her; but in another instant it vanished, and Mrs Cockle found herself mounting — much more slowly now — into a damp, dingy, foggy mist, which was really the heavy rain-cloud she had

noticed earlier in the day. The mist made her feel cold and damp and uncomfortable all over. It was so thick that Mrs Cockle was only rising through it extremely slowly. She began to wonder whether she was going to be the only balloon-seller in London to end her days in a rain-cloud.

Suddenly Mrs Cockle felt a pleasant warmth on her knuckles where, above her head, they grasped the strings of her balloons. She looked up and saw a radiance brightening through the cloud. She was now rising a little faster, and in a few moments she burst out of the rain-cloud altogether. Now she saw why her knuckles had felt warm and why she had seen a brightness: on the other side of the thick clouds the sun was shining warmly and brightly, as on a summer's day. Overhead the sky was a speckless blue, and beneath Mrs Cockle's feet the rain-cloud itself, with the sunshine upon it, now looked as white and gleaming as a snowfield. Mrs Cockle thought it looked rather like the top of her own wedding cake all those years ago — and there was certainly an air of celebration about the glittering scene.

Beneath the clouds, London must be cold and wet and windy, but here the weather was perfect. There was hardly more than a breeze, so that Mrs Cockle's balloons were not tempted to float her up any higher. Instead, they gave her just enough support for her to be able to walk, in a rather feathery, bouncy fashion, on the surface of the cloudland. The feeling of half walking, half floating along, as light as one of her own balloons, was so delightful that Mrs Cockle forgot all her troubles — forgot Peter even. She was a gay old woman at heart, and she had the commonsense to realize that she was unlikely ever to have such an experience as this again. She practised little runs and jumps that, without any effort, carried her for yards at a time. She looked round and admired the changing cloud-scenery — the clouds that moved lazily and puffed themselves up into twisted, toppling mountains, or swirled away to leave mysterious caverns into which she peered, or even drew away altogether to leave gulfs and abysses.

It was while Mrs Cockle was peering over the edge of one of these gaps in the cloud that she suddenly realized how high up she was, and, for the first time, felt giddy. Far, far below were the roofs of London, and no way of getting safely down to them, that she could think of. She wished aloud that she were a cat, like Peter, with nine lives, for she would need them all to get safely to earth again. Then she wished that Peter were there with her, because cats are so resourceful. But, of course, wishing was no good.

Mrs Cockle was a sensible old woman, and resourceful, even without Peter. Through the gap in the clouds, she had caught a glimpse of the River Thames, and now she determined to keep above it as long as she could: she felt that her old bones would fall more softly into water — if fall they must — than on to hard chimney-pots.

So she looked through any holes in the clouds as often as she could, to keep the River Thames in sight, and from above she followed its dark, twisting course as far as she was able. She had to keep on the move, anyway, because, if she stood still, she at once began to sink through the cloud surface, and the damp of the mist would begin to come up over the tops of her goloshes.

In the ordinary way, for someone to walk along the River Thames from London down to the river's mouth would be very long and dull and tiring. For Mrs Cockle, tripping along lightly overhead, it was none of these things. Only towards the end did she even begin to wish there were a cup of tea to be had.

At last, after walking without effort all morning and part of the afternoon, Mrs Cockle came within sight of the end of her cloudland. Really it was not so much that the cloud came to an end, as that it was now beginning to melt away altogether. The bad weather of so many weeks was changing, the clouds were vanishing, and the day was going to clear up. What interested Mrs Cockle, more than the weather, was what lay over the broken edge of the clouds. Soon she saw: it was the sea. She had followed the River Thames from London down to its very mouth, where it runs into the sea. She was over the sea itself.

Her first feeling was of thankfulness that now, if the worst came to the worst, there were no chimney-pots at all to fall upon. Her next feeling was one of curiosity, for Mrs Cockle had never seen the sea — although she had, of course, heard about it. After all, you cannot have a reliable name for selling balloons in London every day in the year *and* be able to take a day off at the seaside. Besides, the train fare would have been more than Mrs Cockle could afford, if she were to keep Peter in the comfort to which he was accustomed. So this was the very first time Mrs Cockle had seen the sea; and she meant to make the most of it.

Her prudence held her back from going too near the cloud edge, but all the same, in her eagerness, she did not notice that the cloud itself was steadily melting away.

Suddenly a hole came under her right foot, which at once sank through it; at the same time, a wisp of cloud curling back on itself tripped up her left foot, and in an instant Mrs Cockle had fallen over the edge. In the confusion of losing her balance Mrs Cockle let go of the balloons, and they soared away into the highest air and were never seen again. For a few seconds after that, Mrs Cockle fell very rapidly; then her enormous umbrella, which had been jerked up without losing hold on her arm, opened out of its own accord — as umbrellas will do, in such circumstances. Mrs Cockle clung tightly to it, and felt it steady her, as it opened above her like a parachute.

Now she was dropping through the air quite gently, although fairly fast. She dared to look down at the wonderful strange sea below, and saw it as she would never see it again — a blue surface scrawled and scribbled over with little wavering lines of white that were foam. Below her, on the sea, was a black speck which, as she came lower, she could make out to be a fishing-boat. It was the first of the boats to put out to sea to take advantage of the better weather. The bottom of the boat was already silver with newly-caught fish, and the young fisherman was already hauling his nets for the second time.

As she came lower, Mrs Cockle called to the man in the boat, but either he was too busy to hear her, or else he could not believe a voice would come from above him. The next minute, Mrs Cockle's feet touched sea-water for the first time in their life, and in the next instant the whole of Mrs Cockle was sinking into it. She could not swim, and she had heard that the sea could be very deep, so that she was relieved but surprised to find something solid — or nearly solid — under her feet. Just then the young fisherman, whose net she was feeling beneath her, called out crossly, 'What are you doing in my fishing net?'

Mrs Cockle would have begun to answer his question but for

something very surprising — the really astounding part of this story. This astounding thing made Mrs Cockle call back, 'And, pray, what are you doing with that cat in your fishing-boat?' For there sat Peter Cockle.

He sat in the stern of the boat, paying no attention to men hauling in nets, or old women falling from the skies into them. He was staring steadily at the fresh fish that were piling up in the bottom of the boat. Although he had been too well brought up by Mrs Cockle not to know how to wait for his meals, the tip of his pink tongue was caught between his teeth as though it had positively to be kept back from the feast.

Peter sat there, while the fisherman hauled his net in and helped Mrs Cockle aboard and said he was sorry if he had spoken roughly but it had all been so unexpected. Mrs Cockle said she was sorry too, and hoped she hadn't done any harm to the nets but she was always very light upon her feet. Then the young man put his dry jacket round Mrs Cockle's shoulders and said he would row her back to land at once, because he could see that what she needed was a cup of hot tea. While he rowed, Mrs Cockle asked him very politely about

the cat who sat so quietly in the stern.

'Oh, him!' said the fisherman. 'He turned up a week or more ago, very thin and raggy-coated, and his foot-pads worn thin with walking from wherever he came from. All the same he looked a handsome cat, so many a one would have taken him in. The milkman tried to tempt him with a dish of cream, but he wouldn't go. Then the grocer tried him with potted shrimps: he'd have none of them. Instead, he spent his time hanging round the fishing boats and nets on the shore. In the end, I believe he took a fancy to me, and I'll tell you why I think it. I've put to sea when other fishermen daren't, and I've always fed him on a little of the catch, however poor it was. It's my belief, ma'am, that that cat is partial to a bit of fresh fish.'

At the end of the fisherman's story, Peter caught Mrs Cockle's eye by mistake. It was no longer any use to pretend that he did not see her, so — looking very ashamed of himself — Peter stepped across and rubbed himself timidly against her goloshes. Mrs Cockle ought to have been very offended, but she loved Peter too dearly. She bent down and tickled him under the chin, at which Peter instantly began to purr. The young fisherman smiled; Mrs Cockle smiled; and Peter

took courage and purred very loudly.

When they reached the land, the fisherman took Peter and Mrs Cockle to the little hut where he lived all by himself. There he made Mrs Cockle a cup of hot, strong tea. When she had finished, she asked the young man if it would be convenient for her to come and keep house for him, since she had taken such a strong fancy to his cat. He was a little surprised at first; but, on turning the idea over in his mind, he saw what an excellent one it was. So things were arranged.

Mrs Cockle settled down to her duties at once. In the mornings she got up very early, cleaned the little house, made the breakfast, packed a lunch for Peter and the fisherman, and said goodbye to them when they sailed away to fish.

Then, if the fancy took her, she went about her old business of selling balloons, which she did now in a sheltered corner of the promenade. She liked the promenade better than her London street: she could see so much of the sky and the sea, and — better still — she could keep an eye on Peter, far out in the fishing-boat.

When she saw the boat coming back, she packed up her balloons and hurried home ahead of the other two, to get their tea ready. There was always a nice piece of fresh fish for Peter's tea, and the young fisherman used to reflect how thoughtful dear Mrs Cockle was of his cat.

Mrs Cockle never told that Peter had once lived with her in London and then left her: she would not have had people think that Peter was light in his affections. She knew in her heart that, after fresh fish for his tea, Peter Cockle valued her company more than anything else in the world.

McBroom and the Big Wind
by Sid Fleischman

I can't deny it — it does get a mite windy out here on the prairie. Why, just last year a blow came ripping across our farm and carried off a pail of sweet milk. The next day it came back for the cow.

But that wasn't the howlin', scowlin', all mighty *big* wind I aim to tell you about. That was just a common little prairie breeze. No account, really. Hardly worth bragging about.

It was the *big* wind that broke my leg. I don't expect you to believe that — yet. I'd best start with some smaller weather and work up to that bone-breaker.

I remember distinctly the first prairie wind that came scampering along after we bought our wonderful one-acre farm. My, that land is rich. Best topsoil in the country. There isn't a thing that won't grow in our rich topsoil, and fast as lightning.

The morning I'm talking about, our oldest boys were helping me to shingle the roof. I had bought a keg of nails, but it turned out those nails were a whit short. We buried them in our wonderful topsoil and watered them down. In five or ten minutes those nails grew a full half-inch.

So there we were, up on the roof, hammering down shingles. There wasn't a cloud in the sky at first. The younger boys were shooting marbles all over the farm and the girls were jumping rope. When I had pounded down the last shingle I said to myself, 'Josh McBroom, that's a mighty stout roof. It'll last a hundred years.'

Just then I felt a small draught on the back of my neck. A moment later one of the girls — it was Polly, as I recall — shouted up to me. 'Pa,' she said, 'do jack rabbits have wings?'

I laughed. 'No, Polly.'

'Then how come there's a flock of jack rabbits flying over the house?'

I looked up. Mercy! Rabbits were flapping their ears across the sky in a perfect V formation, northbound. I knew then we were in for a slight blow.

'Run, everybody!' I shouted to the young 'uns. I didn't want the wind picking them up by the ears. 'Will*jill*hester*chester*peter*polly*tim-tom*mary*larry*andlittle*clarinda* — in the house! Scamper!'

The clothes-line was already beginning to whip around like a jump rope. My dear wife, Melissa, who had been baking a heap of biscuits, threw open the door. In we dashed and not a moment too soon. The wind was snapping at our heels like a pack of wolves. It aimed to barge right in and make itself at home! A prairie wind has no manners at all.

We slammed the door in its teeth. Now, the wind didn't take that politely. It rammed and battered at the door while all of us pushed and shoved to hold the door shut. My, it was a battle! How the house creaked and trembled!

'Push, my lambs,' I yelled. 'Shove!'

At times the door planks bent like barrel staves. But we held that roaring wind out. When it saw there was no getting past us, the zephyr sneaked around the house to the back door. Howsoever, our oldest boy, Will, was too smart for it. He piled Mama's heap of fresh biscuits against the back door. My dear wife, Melissa, is a wonderful cook, but her biscuits *are* terrible heavy. They made a splendid door-stop.

But what worried me most was our wondrous rich topsoil. That thieving wind was apt to make off with it, leaving us with a trifling hole in the ground.

'Shove, my lambs!' I said. 'Push!'

The battle raged on for an hour. Finally the wind gave up butting its fool head against the door. With a great angry sigh it turned and whisked itself away, scattering fence posts as it went.

We all took a deep breath and I opened the door a crack. Hardly a leaf now stirred on the ground. A bird began to twitter. I rushed outside to our poor one-acre farm.

Mercy! What I saw left me pop-eyed. 'Melissa!' I shouted with glee. 'Willjillhesterchesterpeterpollytimtommarylarryandlittleclarinda! Come here, my lambs! Look!'

We all gazed in wonder. Our topsoil was still there — every bit. Bless those youngsters! The boys had left their marbles all over the field, and the marbles had grown as large as boulders. There they sat, huge agates and sparkling glassies, holding down our precious topsoil.

But that rambunctious wind didn't leave empty-handed. It ripped off our new shingle roof. Pulled out the nails, too. We found out later the wind had shingled every burrow in the next county.

Now that was a strong draught. But it wasn't a *big* wind. Nothing like the kind that broke my leg. Still, that prairie gust was an education to me.

'Young 'uns,' I said, after we'd rolled those giant marbles down the hill. 'The next uninvited breeze that comes along, we'll be ready for it. There are two sides to every flapjack. It appears to me the wind can be downright useful on our farm if we let it know who's boss.'

The next gusty day that came along, we put it to work for us. I made a wind plough. I rigged a bed-sheet and tackle to our old farm plough. Soon as a breeze sprung up I'd go tacking to and fro over the farm, ploughing as I went. Our son Chester once ploughed the entire farm in under three minutes.

On Thanksgiving morning Mama told the girls to pluck a large turkey for dinner. They didn't much like that chore, but a prairie gust arrived just in time. The girls stuck the turkey out of the window. The wind plucked that turkey clean, feathers and all.

Oh, we got downright glad to see a blow come along. The young

'uns were always wanting to go out and play in the wind, but Mama was afraid they'd be carried off. So I made them wind-shoes — made 'em out of heavy iron skillets. Out in the breeze those shoes felt light as feathers. The girls would jump rope with the clothes-line. The wind spun the rope, of course.

Many a time I saw the youngsters put on their wind-shoes and go clumping outside with a big tin funnel and all the empty bottles and jugs they could round up. They'd cork the containers jam-full of prairie wind.

Then, come summer, when there wasn't a breath of air, they'd uncork a bottle or two of fresh winter wind and enjoy the cool breeze.

Of course, we had to wind-proof the farm every fall. We'd plant the field in buttercups. My, they were slippery — all that butter, I guess. The wind would slip and slide over the farm without being able to get a purchase of the topsoil. By then the boys and I had re-shingled the roof. We used screws instead of nails.

Mercy! Then came the *big* wind!

It started out gently enough. There were a few jack rabbits and some crows flying backwards through the air. Nothing out of the ordinary.

Of course the girls went outside to jump the clothes-line and the boys got busy laying up bottles of wind for summer. Mama had just baked a batch of fresh biscuits. My, they did smell good! I ate a dozen or so hot out of the oven. And that turned out to be a terrible mistake.

Outside, the wind was picking up ground speed and scattering fence posts as it went.

'Willjillhesterchesterpeterpollytimtommarylarryandlittleclarinda!' I shouted. 'Inside, my lambs. That wind is getting ornery!'

The young 'uns came trooping in and pulled off their wind-shoes. And not a moment too soon. The clothes-line began to whip around

so fast it seemed to disappear. Then we saw a hen-house come flying through the air, with the hens still in it.

The sky was turning dark and mean. The wind came out of the far north, howling and shrieking and shaking the house. In the cupboard, cups chattered in their saucers.

Soon we noticed big balls of fur rolling along the prairie like tumbleweeds. Turned out they were timber wolves from up north. And then an old hollow log came spinning across the farm and split against my chopping-stump. Out rolled a black bear, and was he in a temper! He had been trying to hibernate and didn't take kindly to being awakened. He gave out a roar and looked around for somebody to chase. He saw us at the windows and decided we would do.

The mere sight of him scared the young 'uns and they huddled together, holding hands, near the fireplace.

I got down my shotgun and opened a window. That was a *mistake*! Two things happened at once. The bear was coming on and in my haste I forgot to calculate the direction of the wind. It came shrieking along the side of the house and when I poked the gun-barrel out of

110

the window, well, the wind bent it like an angle iron. That buck-shot flew due south. I found out later it brought down a brace of ducks over Mexico.

But worse than that, when I threw open the window such a draught came in that our young 'uns *were sucked up through the chimney*! Holding hands, they were carried away like a string of sausages.

Mama near fainted away. 'My dear Melissa,' I exclaimed. 'Don't you worry! I'll get our young 'uns back!'

I fetched a rope and rushed outside. I could see the young 'uns up in the sky and blowing south.

I could also see the bear and he could see me. He gave a growl with a mouthful of teeth like rusty nails. He rose up on his hind-legs and came towards me with his eyes glowing red as fire.

I didn't fancy tangling with that monster. I dodged around behind the clothes-line. I kept one eye on the bear and the other on the young 'uns. They were now flying away over the county and hardly looked bigger than Mayflies.

The bear charged towards me. The wind was spinning the clothes-line so fast he couldn't see it. And he charged smack into it. My, didn't he begin to jump! He jumped red-hot pepper, only faster. He had got himself trapped inside the rope and couldn't jump out.

Of course, I didn't lose a moment. I began flapping my arms like a bird. That was such an enormous *big* wind I figured I could fly after the young 'uns. The wind tugged and pulled at me, but it couldn't lift me an inch off the ground.

Tarnation! I had eaten too many biscuits. They were heavy as lead and weighed me down.

The young 'uns were almost out of sight. I rushed to the barn for the wind-plough. Once out in the breeze, the bedsheet filled with wind. Off I shot like a cannon-ball, ploughing a deep furrow as I went.

Didn't I streak along, though! I was making better time than the young 'uns. I kept my hands on the plough handles and steered around barns and farmhouses. I saw hay-stacks explode in the wind. If that wind got any stronger it wouldn't surprise me to see the sun blown off course. It would set in the south at high noon.

I ploughed right along and gained rapidly on the young 'uns. They were still holding hands and just clearing the tree-tops. Before long I was within hailing distance.

'Be brave, my lambs,' I shouted. 'Hold tight!'

I spurted after them until their shadows lay across my path. But the bed-sheet was so swelled out with wind that I couldn't stop the plough. Before I could let go of the handles and jump off I had sailed far *ahead* of the young 'uns.

I heaved the rope into the air.

112

'Willjillhesterchesterpeterpollytimtommarylarryandlittleclarinda!' I shouted as they came flying overhead. 'Hang on!'

Hester missed the rope, and Jill missed the rope, and so did Peter. But Will caught it. I had to dig my heels in the earth to hold them. And then I started back. The young 'uns were too light for the wind. They hung in the air. I had to drag them home on the rope like balloons on a string.

Of course it took most of the day to shoulder my way back through the wind. It was a mighty struggle, I tell you! It was near supper-time when we saw our farmhouse ahead, and that black bear was still jumping rope!

I dragged the young 'uns into the house. The rascals! They had had a jolly time flying through the air, and wanted to do it again! Mama put them to bed with their wind-shoes on.

The wind blew all night, and the next morning that bear was still jumping rope. His tongue was hanging out and he had lost so much weight he was skin and bones.

Finally, about mid-morning, the wind got tired of blowing one way, so it blew the other. We got to feeling sorry for that bear and cut him loose. He was so tuckered out he didn't even growl. He just pointed himself towards the tall timber to find another hollow log to crawl into. But he had lost the fine art of walking. We watched him jump, jump, jump north until he was out of sight.

That was the howlin', scowlin', all mighty *big* wind that broke my leg. It had not only pulled up fence posts, but the *holes* as well. It dropped one of those holes right outside the barn door and I stepped in it.

That's the bottom truth. Everyone on the prairie knows Josh McBroom would rather break his leg than tell a fib.

The Fisherman and the Bottle

by Geraldine McCaughrean

The fisherman was well known hereabouts (said Shahrazad) though I forget his exact name. He used to be a familiar sight on the beach, throwing his net into the surf to catch bass and mullet. He was almost as old as he was poor, but his faith and trust in Allah comforted him.

Arriving at the sea shore and starting to work, he looked at the sky and said:

'O Allah who sends some days red with mullet and others silver with bass and still more black with mud, is it to be a day of the third kind? My net is caught on the bottom, Allah.'

When he finally dragged the net ashore, he found nothing in it but a dead donkey. Moving along the beach, he cast again, and again his net caught on the sea-bed. Looking at the sky he said:

'O Allah who makes fish, donkey, and fisherman, do not grant me the blessing of a second donkey.' He undressed, dived in, and freed the net. This time he pulled in a small mountain of broken clay pots full of black mud.

'O Allah who makes and breaks every man's life, I thank you for this generous gift of broken pottery and mud, but consider how much greater my thanks would be were you to fill my nets with *fish*.'

He cast again, and yet again the net snagged against the rocks.

'O Allah, did my father or mother offend you before I was born, or is this simply Allah's idea of a joke?' said the fisherman, looking at the sky.

But when he hauled the net ashore, this time it contained a rather fine copper bottle. It was green with age, but once emptied and cleaned it would sell for a few dinars in the bazaar. The lead stopper was still in place, and the mouth of the bottle was sealed securely with an elaborate wax seal.

Now the fisherman's education was small and his ignorance was large. He did not recognize the Royal Seal of Suleiman, first and greatest of all Believers and King throughout the empires of Arabia, who lived two thousand years ago. So the fisherman broke the seal with his knife and prised out the lead stopper.

'What nodding-headed man sealed up an empty bottle,' he muttered, 'and threw a thing of value into the sea?'

He shook it vigorously upside down, but only a dust as fine as smoke trickled out: a dust so fine that its weight was less than air. It wreathed upwards from the bottle's neck, changing colour in the light. As dawn expands into daylight and shadow grows into night, so the dust expanded into a vapour. Just as a seed grows into a tree and a second grows into a year, so the vapour grew into a tower of weaving colours as tall as the sky. Then, just as a distant caravan in the heat-haze of a desert becomes little by little distinct, so the floating colours hardened into the leg, the arm, the hand, the head — the body of an immense jinni. The arch of his foot overshadowed the fisherman, and the lowness of the sky forced the creature to bend his head and neck.

A flat, square head overgrown on all sides with reddish stubble; his nose hung like a jug in the centre of a white face pitted and blotched with red. His mouth was as deep as the mines of Africa; the eyes were as yellow as sulphur beds and the purple veins beside them bubbled horribly.

For a moment the fisherman forgot his prayers, his name and all the powers of speech. Amazement stupefied him. Then the sky's beams shook at the sound of the jinni's voice. 'O great Suleiman, defender of Allah, the one true god, I beg forgiveness and will never again — but you are not Suleiman, O smallest worm.'

The fisherman shook his head (as best he was able while leaning over backwards to gape up at the jinni). 'Who let me out of the bottle?'

'I did, sir. Me.'

'In that case, weasel, I have brought you interesting news,' said the jinni, picking a small cloud out of his beard and blowing it away like a dandelion head. 'It should interest you. It touches you closely.'

'News?' squeaked the fisherman. 'Me?'

'News of your death, O smallest and foulest one. Today. Now.'

The fisherman let out a piercing shriek. 'But what have I done?'

'Choose the way I should kill you,' said the jinni without pity, 'but make it horrible or I will think of a more dreadful way.'

The fisherman could only repeat: 'What have I done? What have I done? Pardon me, Allah. Pardon me, O vastest one, but what have I done?'

'Listen, child of a sickly frog,' said the jinni, swatting a flock of birds as they flew past his shoulder. 'I'll tell you my story, but prepare to die when I have told it. I am Sakhr al-Jinni, the ifrit, who rebelled against Suleiman, son of Daud. My army was defeated and my life fell under the foot of King Suleiman.

'How I crept and wept and flattered, until the King, thinking I was sorry, said: "Calm yourself, Sakhr al-Jinni. Promise to obey me and to obey Allah, and I shall forgive you." Forgive? *Me?* Sakhr al-Jinni, Terror of the Lower Hemisphere? I told him: "You will wait a lifetime for my obedience and Allah will wait an eternity before I become a Believer!"

'And so Suleiman stuffed me into this bottle and sealed it with his seal and hurled me into the deep ocean, where I washed and swashed about like a lake squeezed into a cup or a whale squeezed into an egg.'

'But Lord Suleiman
died two thousand years ago!'
exclaimed the fisherman, and the jinni
let out a terrible groan as he remembered
his imprisonment.

'For the first hundred years I swore that
if anyone freed me from that copper bottle I
would grant him three wishes — however greedy.

'But nobody came.

'For the next two hundred years I swore that if
anyone freed me from that copper bottle I would give
him and all his tribe everlasting riches.

'But nobody came.

'For the next five hundred years I swore that if anyone
freed me from that copper bottle I would make him ruler
and owner of all the people of earth!

'But nobody came.

'For the next thousand years I swore . . . and I swore, but
now my oaths were terrible. My patience was gone, my fury
was bigger than the ocean I was floating in. I swore that if
anyone freed me from that copper bottle (*unless, of course,
it was the all-powerful Lord Suleiman*) I would make him the
first to feel the scourge of my revenge! My old enemies are
long since dead. You will have the honour of standing in their
place while I cut you to atoms! I have sworn it.'

With that the jinni drew a cutlass brighter than sheet lightning,
and began warming its edge in the sun's furnace. He had been
looking forward so eagerly for two thousand years to boiling his victim
in terror before mincing him, that he looked down once more to enjoy
the fisherman's despairing face.

The fisherman, however, was looking up at him with one finger
against his nose and one eye winking.

'Tsk, tsk,' said the fisherman. 'Come now, you don't really expect me to believe that, do you?'

The jinni's anger shook the ocean and caused a tidal wave in furthest China. 'The curses of ten thousand dogs on you and all your tribe,' he bellowed. But the fisherman only shook his head and smiled knowingly.

'No, be honest now, where did you come from just now? I know I was shaking out this old bottle when you — when you loomed up, so to speak. But ignorant as I am, I know what can be done and what cannot, and I also know when someone is enjoying a joke at my expense. I couldn't fit one leg into that bottle.' And the fisherman demonstrated, trying unsuccessfully to squeeze one foot through the narrow bottleneck. 'A sheep cut into small pieces would barely fit inside.'

'But I am a *jinni* . . .' said the ifrit, pouting slightly.

'Well *I've* certainly never met a jinni of half your size,' (said the fisherman with perfect honesty), 'who could do such a thing. Take that cutlass for instance — two thousand years in the sea and no rust to show for it? A jinni who wore that sword and climbed inside such a small bottle would surely ruin himself. No, no. It's no good trying to fool me. You may be clever, sir, but it would take true genius to fit a jinni of your magnitude and magnificence into a bottle like this. Speak truly, where did you come from and what put it into your heart to make fun of a poor old fisherman on such a lovely day?'

It seemed that the jinni would burst with frustration. 'Ignorant little worm,' he shouted, shaking his fists and tearing holes in the sky with them. 'Can you not be made to understand the power of Sakhr al-Jinni, Bringer of Death to a thousand such fishermen as you? Watch me! And shake in your rope-soled sandals, for this is the least of my powers and I am more skilled still in killing stupid, ignorant, witless old fishermen!'

Then, just as an old man's sight grows blurred, the body of the jinni broke up into a tower of weaving colours. And just as a rushing river shrinks to a trickle in summer, so the column shrank to a modest fountain of colour. And just as rocks crumble into sand, so the smoke left only a light soot — a fine dust — that trickled back into the bottle.

Snatching up the lead stopper, the fisherman pushed it in on top of the jinni and pressed home the two-thousand-year-old Royal Seal of Lord Suleiman.

'So, Bringer of Death to a thousand fishermen, Terror of the Southern Hemisphere, lie there for another two thousand years. I shall tell my fellow fishermen that this beach is haunted by a hideous ifrit so that they never have my bad luck in drawing up such a monstrous fish. And may Allah send me the quickness of wits, even in the slowness of old age, to put any such jinni as you in its right and proper place!'

So saying, he hurled the bottle as far out to sea as his old arms would let him. And so it was that one humble Believer escaped a cruel and terrible death.

The Selfish Giant
by Oscar Wilde

Every afternoon, as they were coming from school, the children used to go and play in the Giant's garden.

It was a large lovely garden, with soft green grass. Here and there over the grass stood beautiful flowers like stars, and there were twelve peach-trees that in the spring-time broke out into delicate blossoms of pink and pearl, and in the autumn bore rich fruit. The birds sat on the trees and sang so sweetly that the children used to stop their games in order to listen to them. 'How happy we are here!' they cried to each other.

One day the Giant came back. He had been to visit his friend the Cornish ogre, and had stayed with him for seven years. After the seven years were over he had said all that he had to say, for his conversation was limited, and he determined to return to his own castle. When he arrived he saw the children playing in the garden.

'What are you doing here?' he cried in a very gruff voice, and the children ran away.

'My own garden is my own garden,' said the Giant, 'anyone can understand that, and I will allow nobody to play in it but myself.' So he built a high wall all round it, and put up a notice-board.

He was a very selfish Giant.

The poor children had now nowhere to play. They tried to play on the road, but the road was very dusty and full of hard stones, and they did not like it. They used to wander round the high walls when their lessons were over, and talk about the beautiful garden inside. 'How happy we were there!' they said to each other.

Then the Spring came, and all over the country there were little blossoms and little birds. Only in the garden of the Selfish Giant it was still winter. The birds did not care to sing in it as there were no children, and the trees forgot to blossom. Once a beautiful flower put its head out from the grass, but when it saw the notice-board it was so sorry for the children that it slipped back into the ground again, and went off to sleep.

The only people who were pleased were the Snow and the Frost. 'Spring has forgotten this garden,' they cried, 'so we will live here all the year round.' The Snow covered up the grass with her great white cloak, and the Frost painted all the trees silver.

Then they invited the North Wind to stay with them, and he came. He was wrapped in furs, and he roared all day about the garden, and blew the chimney-pots down. 'This is a delightful spot,' he said, 'we must ask the Hail on a visit.' So the Hail came. Every day for three hours he rattled on the roof of the castle till he broke most of the slates, and then he ran round and round the garden as fast as he could go. He was dressed in grey, and his breath was like ice.

'I cannot understand why the Spring is so late in coming,' said the Selfish Giant, as he sat at the window and looked out at his cold, white garden. 'I hope there will be a change in the weather.'

But the Spring never came, nor the Summer. The Autumn gave golden fruit to every garden, but to the Giant's garden she gave none. 'He is too selfish,' she said. So it was always winter there, and the North Wind and the Hail, and the Frost, and the Snow danced about through the trees.

One morning the Giant was lying awake in bed when he heard some lovely music. It sounded so sweet to his ears that he thought it must be the King's musicians passing by. It was really only a little linnet singing outside his window, but it was so long since he had heard a bird sing in his garden that it seemed to him to be the most beautiful music in the world. Then the Hail stopped dancing over his head, and the North Wind ceased roaring, and a delicious perfume came to him through the open casement. 'I believe the Spring has come at last,' said the Giant, and he jumped out of bed and looked out.

What did he see?

He saw a most wonderful sight. Through a little hole in the wall the children had crept in, and they were sitting in the branches of the trees. In every tree that he could see there was a little child. And the trees were so glad to have the children back again that they had covered themselves with blossoms, and were waving their arms gently above the children's heads. The birds were flying about and twittering with delight, and the flowers were looking up through the green grass and laughing.

It was a lovely scene, only in one corner it was still winter. It was the farthest corner of the garden, and in it was standing a little boy. He was so small that he could not reach up to the branches of the tree, and he was wandering all round it, crying bitterly. The poor tree was still covered with frost and snow, and the North Wind was blowing and roaring above it. 'Climb up! little boy,' said the Tree, and it bent its branches down as low as it could, but the boy was too tiny.

And the Giant's heart melted as he looked out. 'How selfish I have been!' he said. 'Now I know why the Spring would not come here. I will put that poor little boy on the top of the tree, and then I will knock down the wall, and my garden shall be the children's playground for ever and ever.' He was really very sorry for what he had done.

So he crept downstairs and opened the front door quite softly, and went out into the garden. But when the children saw him they were so

123

frightened that they all ran away, and the garden became winter again. Only the little boy did not run for his eyes were so full of tears that he did not see the Giant coming. And the Giant stole up behind him and took him gently in his hand, and put him up into the tree. And the tree broke at once into blossom, and the birds came and sang on it, and the little boy stretched out his two arms and flung them round the Giant's neck, and kissed him.

And the other children, when they saw that the Giant was not wicked any longer, came running back, and with them came the Spring. 'It is your garden now, little children,' said the Giant, and he took a great axe and knocked down the wall. And when the people were going to market at twelve o'clock they found the Giant playing with the children in the most beautiful garden they had ever seen.

All day long they played, and in the evening they came to the Giant to bid him good-bye.

'But where is your little companion?' he said. 'The boy I put into the tree.' The Giant loved him the best because he had kissed him.

'We don't know,' answered the children. 'He has gone away.'

'You must tell him to be sure and come tomorrow,' said the Giant. But the children said that they did not know where he lived and had never seen him before, and the Giant felt very sad.

Every afternoon, when school was over, the children came and played with the Giant. But the little boy whom the Giant loved was never seen again. The Giant was very kind to all the children, yet he longed for his first little friend, and often spoke of him. 'How I would like to see him!' he used to say.

Years went over, and the Giant grew very old and feeble. He could not play about any more, so he sat in a huge armchair, and watched the children at their games, and admired his garden. 'I have many beautiful flowers,' he said, 'but the children are the most beautiful flowers of all.'

One winter morning he looked out of his window as he was

dressing. He did not hate the Winter now, for he knew that it was merely the Spring asleep, and that the flowers were resting.

Suddenly he rubbed his eyes in wonder and looked and looked. It certainly was a marvellous sight. In the farthest corner of the garden was a tree quite covered with lovely white blossoms. Its branches were golden, and silver fruit hung down from them, and underneath it stood the little boy he had loved.

Downstairs ran the Giant in great joy, and out into the garden. He hastened across the grass, and came near to the child. And when he came quite close his face grew red with anger, and he said, 'Who hath dared to wound thee?' For on the palms of the child's hands were the prints of two nails, and the prints of two nails were on the little feet.

'Who hath dared to wound thee?' cried the Giant. 'Tell me, that I may take my big sword and slay him.'

'Nay,' answered the child, 'but these are the wounds of Love.'

'Who art thou?' said the Giant, and a strange awe fell on him, and he knelt before the little child.

And the child smiled on the Giant, and said to him, 'You let me play once in your garden, today you shall come with me to my garden, which is Paradise.'

And when the children ran in that afternoon, they found the Giant lying dead under the tree, all covered with white blossoms.

How the Rhinoceros got his Skin

by Rudyard Kipling

Once upon a time, on an uninhabited island on the shores of the Red Sea, there lived a Parsee from whose hat the rays of the sun were reflected in more-than-oriental splendour. And the Parsee lived by the Red Sea with nothing but his hat and his knife and a cooking-stove of the kind that you must particularly never touch. And one day he took flour and water and currants and plums and sugar and things, and made himself one cake which was two feet across and three feet thick. It was indeed a Superior Comestible (*that's* Magic), and he put it on the stove because *he* was allowed to cook on that stove, and he baked it and he baked it till it was all done brown and smelt most sentimental.

But just as he was going to eat it there came down to the beach from the Altogether Uninhabited Interior one Rhinoceros with a horn on his nose, two piggy eyes, and few manners. In those days the Rhinoceros's skin fitted him quite tight. There were no wrinkles in it anywhere. He looked exactly like a Noah's Ark Rhinoceros, but of course much bigger. All the same, he had no manners then, and he has no manners now, and he never will have any manners. He said, 'How!' and the Parsee left that cake and climbed to the top of a palm-tree with nothing on but his hat, from which the rays of the sun were always reflected in more-than-oriental splendour. And the

127

Rhinoceros upset the oil-stove with his nose, and the cake rolled on the sand, and he spiked that cake on the horn of his nose, and he ate it, and he went away, waving his tail, to the desolate and Exclusively Uninhabited Interior which abuts on the islands of Mazanderan, Socotra, and the Promontories of the Larger Equinox.

Then the Parsee came down from his palm-tree and put the stove on its legs and recited the following *Sloka*, which, as you have not heard, I will now proceed to relate:

> Them that takes cakes
> Which the Parsee-man bakes
> Makes dreadful mistakes.

And there was a great deal more in that than you would think.

Because, five weeks later, there was a heatwave in the Red Sea, and everybody took off all the clothes they had. The Parsee took off his hat, but the Rhinoceros took off his skin and carried it over his shoulder as he came down to the beach to bathe. In those days it buttoned underneath with three buttons and looked like a waterproof. He said nothing whatever about the Parsee's cake, because he had eaten it all, and he never had any manners, then, since, or henceforward. He waddled straight into the water and blew bubbles through his nose, leaving his skin on the beach.

Presently the Parsee came by and found the skin, and he smiled one smile that ran all round his face two times. Then he danced three times round the skin and rubbed his hands. Then he went to his camp and filled his hat with cake-crumbs, for the Parsee never ate anything but cake, and never swept out his camp. He took that skin, and he shook that skin, and he scrubbed that skin, and he rubbed that skin just as full of old, dry, stale, tickly cake-crumbs and some burned currants as ever it could *possibly* hold. Then he climbed to the top of his palm-tree and waited for the Rhinoceros to come out of the water and put it on.

And the Rhinoceros did. He buttoned it up with the three buttons, and it tickled like cake-crumbs in bed. Then he wanted to scratch, but that made it worse, and then he lay down on the sands and rolled and rolled and rolled, and every time he rolled the cake-crumbs tickled him worse and worse and worse. Then he ran to the palm-tree and rubbed and rubbed and rubbed himself against it. He rubbed so much and so hard that he rubbed his skin into a great fold over his shoulders, and another fold underneath, where the buttons used to be (but he rubbed the buttons off), and he rubbed some more folds over his legs. And it spoiled his temper, but it didn't make the least difference to the cake-crumbs. They were inside his skin and they tickled. So he went home, very angry indeed and horribly scratchy, and from that day to this every rhinoceros has great folds in his skin and a very bad temper, all on account of the cake-crumbs inside.

But the Parsee came down from his palm-tree, wearing his hat, from which the rays of the sun were reflected in more-than-oriental splendour, packed up his cooking-stove, and went away in the direction of Orotavo, Amygdala, the Upland Meadows of Anantarivo, and the Marshes of Sonaput.

How the Polar Bear Became

by Ted Hughes

When the animals had been on earth for some time they grew tired of admiring the trees, the flowers, and the sun. They began to admire each other. Every animal was eager to be admired, and spent a part of each day making itself look more beautiful.

Soon they began to hold beauty contests.

Sometimes Tiger won the prize, sometimes Eagle, and sometimes Ladybird. Every animal tried hard.

One animal in particular won the prize almost every time. This was Polar Bear.

Polar Bear was white. Not quite snowy white, but much whiter than any of the other creatures. Everyone admired her. In secret, too, everyone was envious of her. But however much they wished that she wasn't quite so beautiful, they couldn't help giving her the prize.

'Polar Bear,' they said, 'with your white fur, you are almost too beautiful.'

All this went to Polar Bear's head. In fact, she became vain. She was always washing and polishing her fur, trying to make it still whiter. After a while she was winning the prize every time. The only times any other creature got a chance to win was when it rained. On those days Polar Bear would say, 'I shall not go out in the wet. The other creatures will be muddy, and my white fur may get splashed.'

Then, perhaps, Frog or Duck would win for a change.

She had a crowd of young admirers who were always hanging around her cave. They were mainly Seals, all very giddy. Whenever she came out they made a loud shrieking roar:

'Ooooooh! How beautiful she is!'

Before long, her white fur was more important to Polar Bear than anything. Whenever a single speck of dust landed on the tip of one hair of it — she was furious.

'How can I be expected to keep beautiful in this country!' she cried then. 'None of you have ever seen me at my best, because of the dirt here. I am really much whiter than any of you have ever seen me. I think I shall have to go into another country. A country where there is none of this dust. Which country would be best?'

She used to talk in this way because then the Seals would cry, 'Oh, please don't leave us. Please don't take your beauty away from us. We will do anything for you.'

And she loved to hear this.

Soon animals were coming from all over the world to look at her. They stared and stared as Polar Bear stretched out on her rock in the sun. Then they went off home and tried to make themselves look like her. But it was no use. They were all the wrong colour. They were black, or brown, or yellow, or ginger, or fawn, or speckled, but not one of them was white. Soon most of them gave up trying to look beautiful. But they still came every day to gaze enviously at Polar Bear. Some brought picnics. They sat in a vast crowd among the trees in front of her cave.

'Just look at her,' said Mother Hippo to her children. 'Now see that you grow up like that.'

But nothing pleased Polar Bear.

'The dust these crowds raise!' she sighed. 'Why can't I ever get away from them? If only there were some spotless, shining country, all for me...'

Now pretty well all the creatures were tired of her being so much more admired than they were. But one creature more so than the rest. He was Peregrine Falcon.

He was a beautiful bird, all right. But he was not white. Time and again, in the beauty contests he was runner-up to Polar Bear.

'If it were not for her,' he raged to himself, 'I should be first every time.'

He thought and thought for a plan to get rid of her. How? How? How? At last he had it.

One day he went up to Polar Bear.

Now Peregrine Falcon had been to every country in the world. He was a great traveller, as all the creatures well knew.

'I know a country,' he said to Polar Bear, 'which is so clean it is even whiter than you are. Yes, yes, I know, you are beautifully white, but this country is even whiter. The rocks are clean glass and the earth is frozen ice-cream. There is no dirt there, no dust, no mud.

You would become whiter than ever in that country. And no one lives there. You could be queen of it.'

Polar Bear tried to hide her excitement.

'I could be queen of it, you say?' she cried. 'This country sounds made for me. No crowds, no dirt? And the rocks, you say, are glass?'

'The rocks,' said Peregrine Falcon, 'are mirrors.'

'Wonderful!' cried Polar Bear.

'And the rain,' he said, 'is white face powder.'

'Better than ever!' she cried. 'How quickly can I be there, away from all these staring crowds and all this dirt?

'I am going to live in another country,' she told the other animals. 'It is too dirty here to live.'

Peregrine Falcon hired Whale to carry his passenger. He sat on Whale's forehead, calling out the directions. Polar Bear sat on the shoulder, gazing at the sea. The Seals, who had begged to go with her, sat on the tail.

After some days, they came to the North Pole, where it is all snow and ice.

'Here you are,' cried Peregrine Falcon. 'Everything just as I said. No crowds, no dirt, nothing but beautiful clean whiteness.'

'And the rocks actually are mirrors!' cried Polar Bear, and she ran to the nearest iceberg to repair her beauty after the long trip.

Every day now, she sat on one iceberg or another, making herself beautiful in the mirror of the ice. Always, near her, sat the Seals. Her fur became whiter and whiter in this new clean country. And as it became whiter, the Seals praised her beauty more and more. When she herself saw the improvement in her looks she said, 'I shall never go back to that dirty old country again.'

And there she is still, with all her admirers around her.

Peregrine Falcon flew back to the other creatures and told them that Polar Bear had gone for ever. They were all very glad, and set about making themselves beautiful at once. Every single one was saying to himself, 'Now that Polar Bear is out of the way, perhaps I shall have a chance of the prize at the beauty contest.'

And Peregrine Falcon was saying to himself, 'Surely, now, I am the most beautiful of all creatures.'

But that first contest was won by Little Brown Mouse for her pink feet.

The Shrinking of Treehorn

by Florence Parry Heide

Something very strange was happening to Treehorn.

The first thing he noticed was that he couldn't reach the shelf in his closet that he had always been able to reach before, the one where he hid his candy bars and bubble gum.

Then he noticed that his clothes were getting too big.

'My trousers are all stretching or something,' said Treehorn to his mother. 'I'm tripping on them all the time.'

'That's too bad, dear,' said his mother, looking into the oven. 'I do hope this cake isn't going to fall,' she said.

'And my sleeves come down way below my hands,' said Treehorn. 'So my shirts must be stretching, too.'

'Think of that,' said Treehorn's mother. 'I just don't know why this cake isn't rising the way it should. Mrs Abernale's cakes are *always* nice. They *always* rise.'

Treehorn started out of the kitchen. He tripped on his trousers, which indeed did seem to be getting longer and longer.

At dinner that night Treehorn's father said, 'Do sit up, Treehorn. I can hardly see your head.'

'I *am* sitting up,' said Treehorn. 'This is as far up as I come. I think I must be shrinking or something.'

'I'm sorry my cake didn't turn out very well,' said Treehorn's mother.

'It's very nice, dear,' said Treehorn's father politely.

By this time Treehorn could hardly see over the top of the table.

'Sit up, dear,' said Treehorn's mother.

'I *am* sitting up,' said Treehorn. 'It's just that I'm shrinking.'

'What, dear?' asked his mother.

'I'm shrinking. Getting smaller,' said Treehorn.

'If you want to pretend you're shrinking, that's all right,' said Treehorn's mother, 'as long as you don't do it at the table.'

'But I *am* shrinking,' said Treehorn.

'Don't argue with your mother, Treehorn,' said Treehorn's father.

'He does look a little smaller,' said Treehorn's mother, looking at Treehorn. 'Maybe he *is* shrinking.'

'Nobody shrinks,' said Treehorn's father.

'Well, I'm shrinking,' said Treehorn. 'Look at me.'

Treehorn's father looked at Treehorn.

'Why, you're shrinking,' said Treehorn's father. 'Look, Emily, Treehorn is shrinking. He's much smaller than he used to be.'

'Oh, dear,' said Treehorn's mother. 'First it was the cake, and now it's this. Everything happens at once.'

'I *thought* I was shrinking,' said Treehorn, and he went into the den to turn on the television set.

Treehorn liked to watch television. Now he lay on his stomach in front of the television set and watched one of his favourite programmes. He had fifty-six favourite programmes.

During the commercials, Treehorn always listened to his mother and father talking together, unless they were having a boring conversation. If they were having a boring conversation, he listened to the commercials.

Now he listened to his mother and father.

'He really is getting smaller,' said Treehorn's mother. 'What will we do? What will people say?'

'Why, they'll say he's getting smaller,' said Treehorn's father. He thought for a moment. 'I wonder if he's doing it on purpose. Just to be different.'

'Why would he want to be different?' asked Treehorn's mother.

Treehorn started listening to the commercial.

The next morning Treehorn was still smaller. His regular clothes were much too big to wear. He rummaged around in his closet until he found some of his last year's clothes. They were much too big, too, but he put them on and rolled up the pants and rolled up the sleeves and went down to breakfast.

Treehorn liked cereal for breakfast. But mostly he liked cereal boxes. He always read every single thing on the cereal box while he was eating breakfast. And he always sent in for the things the cereal box said he could send for.

In a box in his closet Treehorn saved all of the things he had sent in for from cereal box tops. He had puzzles and special rings and flashlights and pictures of all of the presidents and pictures of all the baseball players and he had pictures of scenes suitable for framing, which he had never framed because he didn't like them very much, and he had all kinds of games and pens and models.

Today on the cereal box was a very special offer of a very special whistle that only dogs could hear.

Treehorn did not have a dog, but he thought it would be nice to have a whistle that dogs could hear, even if *he* couldn't hear it. Even if *dogs* couldn't hear it, it would be nice to have a whistle, just to have it.

He decided to eat all of the cereal in the box so he could send in this morning for the whistle. His mother never let him send in for anything until he had eaten all of the cereal in the box.

Treehorn filled in all of the blank spaces for his name and address and then he went to get his money out of the piggy bank on the kitchen counter, but he couldn't reach it.

'I certainly *am* getting smaller,' thought Treehorn. He climbed up on a chair and got the piggy bank and shook out a dime.

His mother was cleaning the refrigerator. 'You know how I hate to have you climb up on the chairs, dear,' she said. She went into the living-room to dust.

Treehorn put the piggy bank in the bottom kitchen drawer.

'That way I can get it no matter *how* little I get,' he thought.

He found an envelope and put a stamp on it and put the dime and the box top in so he could mail the letter on the way to school. The mailbox was right next to the bus stop.

It was hard to walk to the bus stop because his shoes kept slipping off, but he got there in plenty of time, shuffling. He couldn't reach

the mailbox slot to put the letter in, so he handed the letter to one of his friends, Moshie, and asked him to put it in. Moshie put it in. 'How come you can't mail it yourself, stupid?' asked Moshie.

'Because I'm shrinking,' explained Treehorn. 'I'm shrinking and I'm too little to reach the mailbox.'

'That's a stupid thing to do,' said Moshie. 'You're *always* doing stupid things, but that's the *stupidest*.'

When Treehorn tried to get on the school bus, everyone was pushing and shoving. The bus driver said, 'All the way back in the bus, step all the way back.' Then he saw Treehorn trying to climb on to the bus.

'Let that little kid on,' said the driver.

Treehorn was helped on to the bus. The bus driver said, 'You can stay right up here next to me if you want to, because you're so little.'

'It's me, Treehorn,' said Treehorn to his friend the bus driver.

The bus driver looked down at Treehorn. 'You do look like Treehorn, at that,' he said. 'Only smaller. Treehorn isn't that little.'

'I am Treehorn. I'm just getting smaller,' said Treehorn.

'Nobody gets smaller,' said the bus driver. 'You must be Treehorn's kid brother. What's your name?'

'Treehorn,' said Treehorn.

'First time I ever heard of a family naming two boys the same name,' said the bus driver. 'Guess they couldn't think of any other name, once they thought of Treehorn.'

Treehorn said nothing.

When he went into class, his teacher said, 'Nursery school is down at the end of the hall, honey.'

'I'm Treehorn,' said Treehorn.

'If you're Treehorn, why are you so small?' asked the teacher.

'Because I'm shrinking,' said Treehorn. 'I'm getting smaller.'

'Well, I'll let it go for today,' said his teacher. 'But see that it's taken care of before tomorrow. We don't shrink in this class.'

After recess, Treehorn was thirsty, so he went down the hall to the water bubbler. He couldn't reach it, and he tried to jump up high enough. He still couldn't get a drink, but he kept jumping up and down, trying.

His teacher walked by. 'Why, Treehorn,' she said. 'That isn't like you, jumping up and down in the hall. Just because you're shrinking, it does not mean you have special privileges. What if all the children in the *school* started jumping up and down in the halls? I'm afraid you'll have to go to the Principal's office, Treehorn.'

So Treehorn went to the Principal's office.

'I'm supposed to see the Principal,' said Treehorn to the lady in the Principal's outer office.

'It's a very busy day,' said the lady. 'Please check here on this form the reason you have to see him. That will save time. Be sure to put your name down, too. That will save time. And write clearly. That will save time.'

Treehorn looked at the form.

There were many things to check, but Treehorn couldn't find one that said 'Being Too Small to Reach the Water Bubbler'. He finally wrote in 'SHRINKING'.

When the lady said he could see the Principal, Treehorn went into the Principal's office with his form.

The Principal looked at the form, and then he looked at Treehorn. Then he looked at the form again.

'I can't read this,' said the Principal. 'It looks like SHIRKING.

You're not SHIRKING, are you, Treehorn? We can't have any shirkers here, you know. We're a team, and we all have to do our very best.'

'It says SHRINKING,' said Treehorn. 'I'm shrinking.'

'Shrinking, eh?' said the Principal. 'Well, now, I'm very sorry to hear that, Treehorn. You were right to come to me. That's what I'm here for. To guide. Not to punish, but to guide. To guide all the members of my team. To solve all their problems.'

'But I don't have any problems,' said Treehorn, 'I'm just shrinking.'

'Well, I want you to know I'm right here when you need me, Treehorn,' said the Principal, 'and I'm glad I was here to help you. A team is only as good as its coach, eh?'

The Principal stood up. 'Good-bye, Treehorn. If you have any more problems, come straight to me, and I'll help you again. A problem isn't a problem once it's solved, right?'

By the end of the day Treehorn was still smaller.

At the dinner table that night he sat on several cushions so he could be high enough to see over the top of the table.

'He's still shrinking,' sniffed Treehorn's mother. 'Heaven knows I've *tried* to be a good mother.'

'Maybe we should call a doctor,' said Treehorn's father.

'I did,' said Treehorn's mother. 'I called every doctor in the Yellow Pages. But no one knew anything about shrinking problems.'

She sniffed again. 'Maybe he'll just keep getting smaller and smaller until he disappears.'

'No one disappears,' said Treehorn's father positively.

'That's right, they don't,' said Treehorn's mother more cheerfully. 'But no one shrinks, either,' she said after a moment. 'Finish your carrots, Treehorn.'

The next morning Treehorn was so small he had to jump out of bed. On the floor under the bed was a game he'd pushed under there and forgotten about. He walked under the bed to look at it.

It was one of the games he'd sent in for from a cereal box. He had started playing it a couple of days ago, but he hadn't had a chance to finish it because his mother had called him to come right downstairs that minute and have his breakfast or he'd be late for school.

Treehorn looked at the cover of the box.

The game was called THE *BIG* GAME FOR KIDS TO GROW ON.

Treehorn sat under the bed to finish playing the game.

He always liked to finish things, even if they were boring. Even if he was watching a boring programme on TV, he always watched it right to the end. Games were the same way. He'd finish this one now. Where had he left off? He remembered he'd just had to move his piece back seven spaces on the board when his mother had called him.

He was so small now that the only way he could move the spinner was by kicking it, so he kicked it. It stopped at number 4. That meant he could move his piece ahead four spaces on the board.

The only way he could move the piece forward now was by carrying it, so he carried it. It was pretty heavy. He walked along the board to the fourth space. It said: CONGRATULATIONS, AND UP YOU GO: ADVANCE THIRTEEN SPACES.

Treehorn started to carry his piece forward the thirteen spaces, but the piece seemed to be getting smaller. Or else *he* was getting *bigger*. That was it, he *was* getting bigger, because the bottom of the bed was getting close to his head. He pulled the game out from under the bed to finish playing it.

He kept moving the piece forward, but he didn't have to carry it any longer. In fact, he seemed to be getting bigger and bigger with

142

each space he landed in.

'Well, I don't want to get *too* big,' thought Treehorn. So he moved the piece ahead slowly from one space to the next, getting bigger with each space, until he was his own regular size again. Then he put the spinner and the pieces and the instructions and the board back in the box for THE *BIG* GAME FOR KIDS TO GROW ON and put it in his closet. If he ever wanted to get bigger or smaller he could play it again, even if it *was* a pretty boring game.

Treehorn went down for breakfast and started to read the new cereal box. It said you could send for a hundred balloons. His mother was cleaning the living-room. She came into the kitchen to get a dust rag.

'Don't put your elbows on the table while you're eating, dear,' she said.

'Look,' said Treehorn. 'I'm my own size now. My own regular size.'

'That's nice, dear,' said Treehorn's mother. 'It's a very nice size, I'm sure, and if I were you I wouldn't shrink any more. Be sure to tell your father when he comes home tonight. He'll be so pleased.' She went back to the living-room and started to dust and vacuum.

That night Treehorn was watching TV. As he reached over to change channels, he noticed that his hand was bright green. He looked in the mirror that was hanging over the television set. His face was green. His ears were green. His hair was green. He was green all over.

Treehorn sighed. 'I don't think I'll tell anyone,' he thought to himself. 'If I don't say anything, they won't notice.'

Treehorn's mother came in. 'Do turn the volume down a little, dear,' she said. 'Your father and I are having the Smedleys over to play bridge. Do comb your hair before they come, won't you, dear,' said his mother as she walked back to the kitchen.

William's Version
by Jan Mark

William and Granny were left to entertain each other for an hour
while William's mother went to the clinic.

'Sing to me,' said William.

'Granny's too old to sing,' said Granny.

'I'll sing to you, then,' said William. William only knew one song.
He had forgotten the words and the tune, but he sang it several
times, anyway.

'Shall we do something else now?' said Granny.

'Tell me a story,' said William. 'Tell me about the wolf.'

'Red Riding Hood?'

'No, not *that* wolf, the other wolf.'

'Peter and the wolf?' said Granny.

'Mummy's going to have a baby,' said William.

'I know,' said Granny.

William looked suspicious. 'How do you know?'

'Well . . . she told me. And it shows, doesn't it?'

'The lady down the road had a baby. It looks like a pig,' said
William. He counted on his fingers. 'Three babies looks like three
pigs.'

'Ah,' said Granny. 'Once upon a time there were three little pigs.
Their names were —'

144

'They didn't have names,' said William.

'Yes they did. The first pig was called —'

'Pigs don't have names.'

'Some do. These pigs had names.'

'No they didn't.' William slid off Granny's lap and went to open the corner cupboard by the fireplace. Old magazines cascaded out as old magazines do when they have been flung into a cupboard and the door slammed shut. He rooted among them until he found a little book covered with brown paper, climbed into the cupboard, opened the book, closed it and climbed out again. 'They didn't have names,' he said.

'I didn't know you could read,' said Granny, properly impressed.

'C—A—T, wheelbarrow,' said William.

'It that the book Mummy reads to you out of?'

'It's my book,' said William.

'But it's the one Mummy reads?'

'If she says please,' said William.

'Well, that's Mummy's story, then. My pigs have names.'

'They're the wrong pigs.' William was not open to negotiation. 'I don't want them in this story.'

'Can't we have different pigs this time?'

'No. They won't know what to do.'

'Once upon a time,' said Granny, 'there were three little pigs who lived with their mother.'

'Their mother was dead,' said William.

'Oh, I'm sure she wasn't,' said Granny.

'She was dead. You make bacon out of dead pigs. She got eaten for breakfast and they threw the rind out for the birds.'

'So the three little pigs had to find homes for themselves.'

'No.' William consulted his book. 'They had to build little houses.'

'I'm just coming to that.'

145

'You said they had to *find* homes. They didn't *find* them.'

'The first little pig walked along for a bit until he met a man with a load of hay.'

'It was a lady.'

'A lady with a load of hay?'

'NO! It was a lady-pig. You said he.'

'I thought all the pigs were little boy-pigs,' said Granny.

'It says lady-pig here,' said William. 'It says the lady-pig went for a walk and met a man with a load of hay.'

'So the lady-pig,' said Granny, 'said to the man, "May I have some of that hay to build a house?" and the man said, "Yes." Is that right?'

'Yes,' said William. 'You know that baby?'

'What baby?'

'The one Mummy's going to have. Will that baby have shoes on when it comes out?'

'I don't think so,' said Granny.

'It will have cold feet,' said William.

'Oh no,' said Granny. 'Mummy will wrap it up in a soft shawl, all snug.'

'I don't *mind* if it has cold feet,' William explained. 'Go on about the lady-pig.'

'So the little lady-pig took the hay and built a little house. Soon the wolf came along and the wolf said —'

'You didn't tell where the wolf lived.'

'I don't know where the wolf lived.'

'15 Tennyson Avenue, next to the bomb-site,' said William.

'I bet it doesn't say that in the book,' said Granny, with spirit.

'Yes it does.'

'Let me see, then.'

William folded himself up with his back to Granny, and pushed the book up under his pullover.

'*I* don't think it says that in the book,' said Granny.

'It's in ever so small words,' said William.

'So the wolf said, "Little pig, little pig, let me come in," and the little pig answered, "No". So the wolf said, "Then I'll huff and I'll puff and I'll blow your house down," and he huffed and he puffed and he blew the house down, and the little pig ran away.'

'He ate the little pig,' said William.

'No, no,' said Granny. 'The little pig ran away.'

'He ate the little pig. He ate her in a sandwich.'

'All right, he ate the little pig in a sandwich. So the second little pig —'

'You didn't tell about the tricycle.'

'What about the tricycle?'

'The wolf got on his tricycle and went to the bread shop to buy some bread. To make the sandwich,' William explained, patiently.

'Oh well, the wolf got on his tricycle and went to the bread shop to buy some bread. And he went to the grocer's to buy some butter.' This innovation did not go down well.

'He already had some butter in the cupboard,' said William.

'So then the second little pig went for a walk and met a man with a load of wood, and the little pig said to the man, "May I have some of that wood to build a house?" and the man said, "Yes."'

'He didn't say please.'

'"Please may I have some of that wood to build a house?"'

'It was sticks.'

'Sticks *are* wood.'

William took out his book and turned the pages. 'That's right,' he said.

'Why don't you tell the story?' said Granny.

'I can't remember it,' said William.

'You could read it out of your book.'

'I've lost it,' said William, clutching his pullover.

'Look, do you know who this is?' He pulled a green angora scarf from under the sofa.

'No, who is it?' said Granny, glad of the diversion.

'This is Doctor Snake.' He made the scarf wriggle across the carpet.

'Why is he a doctor?'

'Because he is all furry,' said William. He wrapped the doctor round his neck and sat sucking the loose end. 'Go on about the wolf.'

'So the little pig built a house of sticks and along came the wolf — on his tricycle?'

'He came by bus. He didn't have any money for a ticket so he ate up the conductor.'

'That wasn't very nice of him,' said Granny.

'No,' said William. 'It wasn't *very* nice.'

'And the wolf said, "Little pig, little pig, let me come in," and the little pig said, "No," and the wolf said, "Then I'll huff and I'll puff and I'll blow your house down," so he huffed and he puffed and he blew the house down. And then what did he do?' Granny asked, cautiously.

William was silent.

'Did he eat the second little pig?'

'Yes.'

'How did he eat this little pig?' said Granny, prepared for more pig

148

sandwiches or possibly pig on toast.

'With his mouth,' said William.

'Now the third little pig went for a walk and met a man with a load of bricks. And the little pig said, "*Please* may I have some of those bricks to build a house?" and the man said, "Yes." So the little pig took the bricks and built a house.'

'He built it on the bomb-site.'

'Next door to the wolf?' said Granny. 'That was very silly of him.'

'There wasn't anywhere else,' said William. 'All the roads were full up.'

'The wolf didn't have to come by bus or tricycle this time, then, did he?' said Granny, grown cunning.

'Yes.' William took out the book and peered in, secretively. 'He was playing in the cemetery. He had to get another bus.'

'And did he eat the conductor this time?'

'No. A nice man gave him some money, so he bought a ticket.'

'I'm glad to hear it,' said Granny.

'He ate the nice man,' said William.

'So the wolf got off the bus and went up to the little pig's house, and he said, "Little pig, little pig, let me come in," and the little pig said, "No," and then the wolf said, "I'll huff and I'll puff and I'll blow your house down," and he huffed and he puffed and he huffed and he puffed but he couldn't blow the house down because it was made of bricks.'

'He couldn't blow it down,' said William, 'because it was stuck to the ground.'

'Well, anyway, the wolf got very

cross then, and he climbed on the roof and shouted down the chimney, "I'm coming to get you!" but the little pig just laughed and put a big saucepan of water on the fire.'

'He put it on the gas stove.'

'He put it on the *fire*,' said Granny, speaking very rapidly, 'and the wolf fell down the chimney and into the pan of water and was boiled and the little pig ate him for supper.'

William threw himself full length on the carpet and screamed.

'He didn't! He didn't! *He didn't!* He didn't eat the wolf.'

Granny picked him up, all stiff and kicking, and sat him on her lap.

'Did I get it wrong again, love? Don't cry. Tell me what really happened.'

William wept, and wiped his nose on Doctor Snake.

'The little pig put the saucepan on the gas stove and the wolf got down the chimney and put the little pig in the saucepan and boiled him. He had him for tea, with chips,' said William.

'Oh,' said Granny. 'I've got it all wrong, haven't I? Can I see the book, then I shall know, next time.'

William took the book from under his pullover. Granny opened it and read, *First Aid for Beginners: a Practical Handbook.*

'I see,' said Granny. 'I don't think I can read this. I left my glasses at home. You tell Gran how it ends.'

William turned to the last page which showed a prostrate man with his leg in a splint; *compound fracture of the femur.*

'Then the wolf washed up and got on his tricycle and went to see his Granny, and his Granny opened the door and said, "Hello, William."'

'I thought it was the wolf.'

'It was. It was the wolf. His name was William Wolf,' said William.

'What a nice story,' said Granny. 'You tell it much better than I do.'

150

Paul's Tale

by Mary Norton

' "Ho! Ho!" said the King, slapping his fat thighs. "Methinks this youth shows promise." But at that moment the Court Magician stepped forward…' 'What is the matter, Paul? Don't you like this story?'

'Yes, I like it.'

'Then lie quiet, dear, and listen.'

'It was just a sort of stalk of a feather pushing itself up through the eiderdown.'

'Well, don't help it, dear, it's destructive. Where were we?' Aunt Isobel's short-sighted eyes searched down the page of the book: she looked comfortable and pink, rocking there in the firelight . . .

' "stepped forward . . . You see the Court Magician knew that the witch had taken the magic music-box, and that Colin . . ." Paul, you aren't listening!'

'Yes, I am. I can hear.'

'Of course you can't — right under the bed-clothes. What are you doing, dear?'

'I'm seeing what a hot water bottle feels like.'

'Don't you know what a hot water bottle feels like?'

'I know what it feels like to me. I don't know what it feels like to itself.'

'Well, shall I go on or not?'

'Yes, go on,' said Paul. He emerged from the bed-clothes, his hair ruffled.

Aunt Isobel looked at him curiously. He was her godson; he had a bad feverish cold; his mother had gone to London. 'Does it tire you, dear, to be read to?' she said at last.

'No. But I like told stories better than read stories.'

Aunt Isobel got up and put some more coal on the fire. Then she looked at the clock. She sighed. 'Well, dear,' she said brightly, as she sat down once more on the rocking-chair. 'What sort of story would you like?' She unfolded her knitting.

'I'd like a real story.'

'How do you mean, dear?' Aunt Isobel began to cast on. The cord of her pince-nez, anchored to her bosom, rose and fell in gentle undulations.

Paul flung round on his back, staring at the ceiling. 'You know,' he said, 'quite real — so you know it must have happened.'

'Shall I tell you about Grace Darling?'

'No, tell me about a little man.'

'What sort of little man?'

'A little man just as high —' Paul's eyes searched the room — 'as that candlestick on the mantelshelf, but without the candle.'

'But that's a very small candlestick. It's only about six inches.'

'Well, about that big.'

Aunt Isobel began knitting a few stitches. She was disappointed about the fairy story. She had been reading with so much expression, making a deep voice for the king, and a wicked oily voice for the Court Magician, and a fine cheerful boyish voice for Colin, the swineherd. A little man — what could she say about a little man? 'Ah!' she exclaimed suddenly, and laid down her knitting, smiling at Paul. 'Little men . . . of course . . .

'Well,' said Aunt Isobel, drawing in her breath. 'Once upon a time there was a little, tiny man, and he was no bigger than that candlestick — there on the mantelshelf.'

Paul settled down, his cheek on his crook'd arm, his eyes on Aunt Isobel's face. The firelight flickered softly on the walls and ceiling.

'He was the sweetest little man you ever saw, and he wore a little red jerkin and a dear little cap made out of a foxglove. His boots . . .'

'He didn't have any,' said Paul.

Aunt Isobel looked startled. 'Yes,' she exclaimed. 'He had boots — little, pointed —'

'He didn't have any clothes,' contradicted Paul. 'He was bare.'

Aunt Isobel looked perturbed. 'But he would have been cold,' she pointed out.

'He had thick skin,' explained Paul. 'Like a twig.'

'Like a twig?'

'Yes. You know that sort of wrinkly, nubbly skin on a twig.'

Aunt Isobel knitted in silence for a second or two. She didn't like the little naked man nearly as much as the little clothed man: she was trying to get used to him. After a while she went on.

'He lived in a bluebell wood, among the roots of a dear old tree. He had a dear little house, tunnelled out of the soft, loamy earth, with a bright blue front door.'

'Why didn't he live in it?' asked Paul.

'He did live in it, dear,' explained Aunt Isobel patiently.

'I thought he lived in the potting-shed.'

'In the potting-shed?'

'Well, perhaps he had two houses. Some people do. I wish I'd seen the one with the blue front door.'

'Did you see the one in the potting-shed?' asked Aunt Isobel, after a moment's silence.

'Not inside. Right inside. I'm too big. I just sort of saw into it with a flashlight.'

'And what was it like?' asked Aunt Isobel, in spite of herself.

153

'Well, it was clean — in a potting-shed sort of way. He'd made the furniture himself. The floor was just earth but he'd trodden it down so that it was hard. It took him years.'

'Well, dear, you seem to know more about this little man than I do.'

Paul snuggled his head more comfortably against his elbow. He half-closed his eyes. 'Go on,' he said dreamily.

Aunt Isobel glanced at him hesitatingly. How beautiful he looked, she thought, lying there in the firelight with one curled hand lying lightly on the counterpane. 'Well,' she went on, 'this little man had a little pipe made out of straw.' She paused, rather pleased with this idea. 'A little hollow straw, through which he played jiggity little tunes. And to which he danced.' She hesitated. 'Among the bluebells,' she added. Really this was quite a pretty story. She knitted hard for a few seconds, breathing heavily, before the next bit would come. 'Now,' she continued brightly, in a changed, higher and more conversational voice, 'up the tree, there lived a fairy.'

'In the tree?' asked Paul, incredulously.

'Yes,' said Aunt Isobel, 'in the tree.'

154

Paul raised his head. 'Do you know this for certain?'

'Well, Paul,' began Aunt Isobel. Then she added playfully, 'Well, I suppose I do.'

'Go on,' said Paul.

'Well, this fairy . . .'

Paul raised his head again. 'Couldn't you go on about the little man?'

'But dear, we've done the little man — how he lived in the roots, and played a pipe, and all that.'

'You didn't say about his hands and feet.'

'His hands and feet!'

'How sort of big his hands and feet looked, and how he could scuttle along. Like a rat,' Paul added.

'Like a rat!' exclaimed Aunt Isobel.

'And his voice. You didn't say anything about his voice.'

'What sort of voice,' Aunt Isobel looked almost scared, 'did he have?'

'A croaky sort of a voice. Like a frog. And he says "Will'ee" and "Doo'ee".'

'Willy and Dooey . . .' repeated Aunt Isobel.

'Instead of "Will you" and "Do you". You know.'

'Has he got a Sussex accent?'

'Sort of. He isn't used to talking. He is the last one. He's been all alone, for years and years.'

'Did he —' Aunt Isobel swallowed. 'Did he tell you that?'

'Yes. He had an aunt and she died about fifteen years ago. But even when she was alive, he never spoke to her.'

'Why?' asked Aunt Isobel.

'He didn't like her,' said Paul.

There was silence. Paul stared dreamily into the fire. Aunt Isobel sat as if turned to stone, her hands idle in her lap. After a while, she cleared her throat.

'When did you first see this little man, Paul?'

'Oh, ages and ages ago. When did you?'

'I — Where did you find him?'

'Under the chicken house.'

'Did you — did you speak to him?'

Paul made a little snort. 'No. I just popped a tin over him.'

'You caught him!'

'Yes. There was an old, rusty chicken-food tin near. I just popped it over him.' Paul laughed. 'He scrabbled away inside. Then I popped an old kitchen plate that was there on top of the tin.'

Aunt Isobel sat staring at Paul. 'What — what did you do with him then?'

'I put him in a cake-tin, and made holes in the lid. I gave him a bit of bread and milk.'

'Didn't he — say anything?'

'Well, he was sort of croaking.'

'And then?'

'Well, I sort of forgot I had him.'

'You forgot!'

'I went fishing, you see. Then it was bedtime. And next day I didn't remember him. Then when I went to look for him, he was lying curled up at the bottom of the tin. He'd gone all soft. He just hung over my finger. All soft.'

Aunt Isobel's eyes protruded dully.

'What did you do then?'

'I gave him some cherry cordial in a fountain-pen filler.'

'That revived him?'

'Yes, that's when he began to talk. And he told me all about his aunt and everything. I harnessed him up, with a bit of string.'

'Oh, Paul,' exclaimed Aunt Isobel, 'how cruel.'

'Well, he'd have got away. It didn't hurt him. Then I tamed him.'

'How did you tame him?'

'Oh, how do you tame anything. With food mostly. Chips of gelatine and raw sago he liked best. Cheese, he liked. I'd take him

156

out and let him go down rabbit holes and things on the string. Then he would come back and tell me what was going on. I put him down all kinds of holes in trees and things.'

'Whatever for?'

'Just to know what was going on. I have all kinds of uses for him.'

'Why,' stammered Aunt Isobel, half-rising from her chair, 'you haven't still got him, have you?'

Paul sat up on his elbows. 'Yes. I've got him. I'm going to keep him till I go to school. I'll need him at school like anything.'

'But it isn't — you wouldn't be allowed —' Aunt Isobel suddenly became extremely grave. 'Where is he now?'

'In the cake-tin.'

'Where is the cake-tin?'

'Over there. In the toy cupboard.'

Aunt Isobel looked fearfully across the shadowed room. She stood up. 'I am going to put the light on, and I shall take that cake-tin out into the garden.'

'It's raining,' Paul reminded her.

'I can't help that,' said Aunt Isobel. 'It is wrong and wicked to keep a little thing like that, shut up in a cake-tin. I shall take it out on to the back porch and open the lid.'

'He can hear you,' said Paul.

'I don't care if he can hear me.' Aunt Isobel walked towards the door. 'I'm thinking of his good, as much as of anyone else's.' She switched on the light. 'Now, which was the cupboard?'

'That one, near the fireplace.'

The door was ajar. Timidly Aunt Isobel pulled it open with one finger. There stood the cake-tin amid a medley of torn cardboard, playing cards, pieces of jig-saw puzzle and an open paint box.

'What a mess, Paul!'

Nervously Aunt Isobel stared at the cake-tin. The holes in the lid were narrow and wedge-shaped, made, no doubt, by the big blade of

the best cutting-out scissors. Aunt Isobel drew in her breath sharply. 'If you weren't ill, I'd make you do this. I'd make you carry out the tin and watch you open the lid —' She hesitated as if unnerved by the stillness of the rain-darkened room and the sinister quiet within the cake-tin.

Then, bravely, she put out a hand. Paul watched her absorbed, as she stretched forward the other one and, very gingerly, picked up the cake-tin. His eyes were dark and deep. He saw the lid was not quite on. He saw the corner, in contact with that ample bosom, rise. He saw the sharp edge catch the cord of Aunt Isobel's pince-nez and, fearing for her rimless glasses, he sat up in bed.

Aunt Isobel felt the tension, the pressure of the glasses on the bridge of her nose. A pull it was, a little steady pull as if a small dark claw, as wrinkled as a twig, had caught the hanging cord . . .

'Look out!' cried Paul.

Loudly she shrieked and dropped the tin. It bounced away and then lay still, gaping emptily on its side. In the horrible hush, they heard the measured clanking of the lid as it trundled off beneath the bed.

Paul broke the silence with a croupy cough.

'Did you see him?' he asked, hoarse but interested.

'No,' stammered Aunt Isobel, almost with a sob. 'I didn't. I didn't see him.'

'But you nearly did.'

Aunt Isobel sat down limply in the upholstered chair. Her hand wavered vaguely round her brow and her cheeks looked white and pendulous, as if deflated. 'Yes,' she muttered, shivering slightly. 'Heaven help me — I nearly did.'

Paul gazed at her a moment longer. 'That's what I mean,' he said.

'What?' asked Aunt Isobel weakly, but as if she did not really care.

Paul lay down again. Gently, sleepily, he pressed his face into the pillow.

'About stories. Being real . . .'

ACKNOWLEDGEMENTS AND SUGGESTIONS FOR FURTHER READING

Joan AIKEN: 'Think of a Word' is from *Hundreds and Hundreds*, ed. Peter Dickinson (Puffin/Viking) © Joan Aiken. Other books of short stories by Joan Aiken include *Past Eight O'Clock*, *Cuckoo Tree*, *Necklace of Raindrops*, and *Tale of a One-Way Street* (all Puffin). **Hans Christian ANDERSEN**: 'The Tinderbox' is from *The Complete Fairy Tales and Stories of Hans Andersen*. Reprinted by permission of Doubleday and Victor Gollancz. This also includes stories such as 'The Ugly Duckling', 'The Emperor's New Clothes', 'Thumbelina', 'The Little Mermaid', and 'The Snow Queen'. **Leila BERG**: 'Pete and the Ladybird' is from *Little Pete Stories* (Methuen). Reprinted with permission. This also includes stories such as 'Pete and the Steam-Roller', 'Pete and the Letter', 'Pete and the Whistle', and 'Pete and the Wonderful Tap'. Also by Leila Berg is *My Dog Sunday* (Puffin). **James BERRY**: 'Bro Tiger Goes Dead' is from *Anancy Spiderman* (Walker Books). Reprinted by permission of Walker Books Ltd. and Henry Holt & Co. Inc. This also includes stories such as 'Anancy, Old Witch and King-Daughter', 'Monkey, Tiger and the Magic Trials', 'Tiger and the Stump-a-Foot Celebration Dance', and 'Anancy Runs into Tiger's Trouble'. **Godfried BOMANS**: 'The Thrush Girl' is from *The Wily Wizard and the Wicked Witch* (1969), translated by Patricia Crampton. Reprinted by permission of J M Dent & Sons. **George BROWNE**: 'The Wonderful Tar Baby' is from *Brer Rabbit Stories* (Jackanory, BBC). Reprinted by permission of BBC Enterprises Ltd. Other stories include 'How Brer Rabbit Got His Tail', 'The Moon in the Mill Pond', and 'Brer Rabbit Saves His Family'. **Kevin CROSSLEY-HOLLAND**: 'The Dauntless Girl' is from *The Dead Moon and Other Tales*. Reprinted by permission of Scholastic Publications Ltd. Other collections of stories by Kevin Crossley-Holland include *Axe Age, Wolf Age*, and *The Faber Book of Northern Folk Tales*. **Dorothy EDWARDS**: 'My Naughty Little Sister Makes a Bottle-Tree' is from *My Naughty Little Sister*. Reprinted by permission of Methuen Children's Books and published in the USA by Clarion Books. Other books of stories about 'My Naughty Little Sister' include *More Naughty Little Sister Stories*, *My Naughty Little Sister and Bad Harry*, *My Naughty Little Sister's Friends*, and *When My Naughty Little Sister was Good*. **Sid FLEISCHMAN**: 'McBroom and the Big Wind' is from *McBroom's Wonderful One Acre Farm* © A.S. Fleischman 1967. Reprinted by permission of Random House UK Ltd. and Curtis Brown Associates. Another book of McBroom stories is *Here Comes McBroom* (Puffin); further stories by Sid Fleischman are *Whipping Boy* (Cornerstone Books) and *Man on the Moon-Eyed Horse* (Gollancz). **David HARRISON**: 'The Giant Who Threw Tantrums' is from *The Book of Giant Stories* (Cape). Reprinted by permission of the author. This also includes 'The Little Boy's Secret' and 'The Giant Who Was Afraid of Butterflies'. **Florence Parry HEIDE**: *The Shrinking of Treehorn* is published by Puffin. © Florence Parry Heide 1971. Reprinted by permission of Penguin Books Ltd. and Holiday House Inc. Other stories about Treehorn are *Treehorn's Treasure* (Puffin) and *Treehorn's Wish* (OUP). **Ted HUGHES**: 'How the Polar Bear Became' is from *How the Whale Became and Other Stories*. Reprinted by permission of Faber and Faber Ltd. Other stories by Ted Hughes include *The Earth-Owl and Other Moon-People*, *The Iron Man*, and *Nessie the Mannerless Monster*. **Terry JONES**: 'Simple Peter's Mirror' is from *Fairy Tales* (Pavilion/Michael Joseph). Reprinted by permission of Terry Jones and Pavilion Books. This also includes stories such as 'The Wonderful Cake-Horse', 'The Butterfly Who Sang', 'The Witch and the Rainbow Cat', 'The Man Who Owned the Earth', and 'The Beast with a Thousand Teeth'. **Rudyard KIPLING**: 'How the Rhinoceros got his Skin' is from *Just So Stories* (Macmillan). This also includes such stories as 'How the Camel Got His Hump', 'How the Leopard Got his Spots', 'How the Alphabet Was Made', and 'The Butterfly that Stamped'. **Julius LESTER**: 'Knee-High Man' is from *Knee-High Man*. Reprinted by permission of Penguin USA. Other stories by Julius Lester include *Mr Rabbit and Mr Bear* (BBC), *Taste of Freedom* (Longman), and *Two Love Stories* (Viking). **Arnold LOBEL**: 'The List' is from *Frog and Toad Together*. Reprinted by permission of William Heinemann and published in the USA by HarperCollins. Other books of Frog and Toad stories are *Frog and Toad All Year*, *Days with Frog and Toad*, and *Frog and Toad Are Friends* (all Puffin). **Geraldine McCAUGHREAN**: 'The Fisherman and the Bottle' is from *One Thousand and One Arabian Nights* (OUP). Reprinted by permission of the author. This also includes stories such as 'The Everlasting Shoes', 'The Wonderful Story of Ali Baba and the Forty Bandits', 'The Prince and the Large and Lonely Tortoise', 'The Ass and his Ass', and 'The Tale of Ala-al Din and his Wonderful Lamp'. **Ruth MANNING-SANDERS:** 'The Fish Cart' is from *Fox Tales* (Methuen). Reprinted by permission of David Higham Associates Ltd. Other books of stories by Ruth Manning-Sanders include *Book of Dwarfs*, *Book of Giants*, *Book of Monsters*, and *Choice of Magic* (all Methuen). **Margaret MAHY:** *The Lion in the Meadow* © Margaret Mahy 1969, 1986. Reprinted by permission of J.M. Dent & Sons Ltd. and The Overlook Press. Other books by Margaret Mahy include *The Boy Who Bounced and Other Stories*, *Chocolate Porridge and Other Stories*, *The Door in the Air and Other Stories*, and *Downhill Crocodile Whizz* (all Puffin). **Jan MARK:** 'William's Version' comes from Jan Mark's book of stories called *Nothing To Be Afraid Of* (Kestrel Books), © Jan Mark 1957. Reprinted by permission of Penguin Books Ltd. and Murray Pollinger. This has other stories you will enjoy. She has written many other books including *The Deadletter Box*. We specially recommend her picture book *Strat and Chatto* in which a clever rat outwits the cat. **Jill McDONALD:** *Maggy Scraggle Loves the Beautiful Ice-Cream Man* is published by Puffin, © Jill McDonald 1978. Reprinted with permission. **Mary NORTON:** 'Paul's Tale' is from *Listening and Writing* (BBC), © Mary Norton. Other stories by Mary Norton include *Bedknobs and Broomsticks*, *The Borrowers*, *The Borrowers Afloat* and *The Borrowers Afield* (all Puffin). **Philippa PEARCE:** *Mrs Cockle's Cat* is published by Viking/Kestrel. Reprinted with permission. Other books by Philippa Pearce include *Battle of Bubble and Squeak*, *A Dog So Small*, *The Squirrel Wife*, and *Who's Afraid and Other Strange Stories* (all Puffin). **Catherine STORR:** 'The Riddlemaster' is from *The Adventures of Polly and the Wolf* (Faber). Reprinted with permission. Other books of stories about Polly and the Wolf include *Clever Polly and the Stupid Wolf*, *Tales of Polly and the Hungry Wolf*, and *Polly and the Wolf Again* (all Puffin). **Oscar WILDE:** 'The Selfish Giant' is from *The Happy Prince and Other Stories* (Puffin). Other stories include 'The Remarkable Rocket', 'The Nightingale and the Rose', 'The Young King', and 'The Fisherman and His Soul'. **Richard WILSON:** 'The Laughing Dragon' is from *The Youngest Omnibus* (Nelson).

Although every effort has been made to secure copyright permission prior to publication, this has not proved possible in some instances. If notified, the publisher will be pleased to rectify any errors or omissions at the earliest opportunity.

THE ARTISTS

Anthony Lewis: A Lion in the Meadow, The Knee-High Man, Bro Tiger Goes Dead, The Dauntless Girl, The Fisherman and the Bottle, William's Version. Illustrations © Anthony Lewis 1994.

Toni Goffe: Maggy Scraggle Loves the Beautiful Ice-Cream Man, The Laughing Dragon, Think of a Word, How the Rhinoceros got his Skin.

Claudio Muñoz: Pete and the Ladybird, My Naughty Little Sister Makes a Bottle-Tree, Simple Peter's Mirror, Mrs Cockle's Cat, The Selfish Giant, Paul's Tale.

Emma Chichester Clark: A List, The Thrush Girl, The Tinderbox.

Elke Counsell: The Fish Cart, The Wonderful Tar Baby, The Riddlemaster, McBroom and the Big Wind, How the Polar Bear Became.

Jean Claverie: The Giant Who Threw Tantrums, The Shrinking of Treehorn.

The jacket illustration and the illustration on the title page are by Anthony Lewis. The illustrations on the contents page are by Toni Goffe and the endpapers are by Claudio Muñoz. We are grateful to Wheatley Primary School for helping us with ideas for the jacket.

Oxford University Press, Walton Street, Oxford OX2 6DP. *Oxford New York Toronto Delhi Bombay Calcutta Madras Karachi Kuala Lumpur Singapore Hong Kong Tokyo Nairobi Dar es Salaam Cape Town Melbourne Auckland Madrid* and associated companies in *Berlin Ibadan*. *Oxford* is a trade mark of Oxford University Press. This selection and arrangement © Michael Harrison and Christopher Stuart-Clark 1994. First published in 1994. All rights reserved. A CIP catalogue record for this book is available from the British Library. ISBN 0 19 278133 2. Printed in Hong Kong.   40-574-1